Estimating and tendering for building work

R. C. Smith

BA, ARICS

Estimating and Tendering for building work

Longman Scientific & Technical

Longman Scientific & Technical
Longman Group UK Limited,
Longman House, Burnt Mill, Harlow,
Essex CM20 2JE, England
and Associated Companies throughout the world

First published 1986
Fifth impression 1992
Sixth impression 1994

British Library Cataloguing in Publication Data
Smith, R. C.
Tenders and estimating for construction work.
1. Building – Estimates – Great Britain
I. Title
692'5'0941 TH 435

ISBN 0582-41173-4

Produced through Longman Malaysia, ACM

Contents

Acknowledgments

We are grateful to the following for permission to reproduce copyright material:
the authors' agents for an extract from *I Could have Kicked Myself* by David Frost & Michael Deakin, Copyright (c) by David Paradine Productions Ltd and the Sunday Express Magazine for the cartoon that accompanied this extract, our Fig. 4.2; the Controller of Her Majesty's Stationery Office for Fig 1 from *BRE Digest* 247, Crown Copyright; Standing Joint Committee for the Standard Method of Measurement of Building Works for extracts from *SMM* 6th edition.

Introduction

The subject of estimating revolves principally around the need for all interested parties involved in the building process, but particularly the Contractor, to be able to predict as accurately as possible the costs of construction.

Traditionally in the construction industry but unlike many other industries, contractors have been required to commit themselves to a price *before* carrying out any work on site and in most cases many months or even years before the work is completed. This makes estimating a very risky occupation, although it could be said that this risk is now mitigated to some extent by the more widespread use of 'fluctuation' contracts.

The estimator

The person in the contractor's organisation responsible for the preparation of prices and estimates for building work is the **Estimator**, and it is he who must be able to use his skill, experience and judgement in attempting to assess the extent of likely future costs. Dictionary definitions of estimating include phrases such as 'approximate judgement of . . .', 'forecasts of . . .', 'form an opinion of . . .', and even 'a prophesy of . . .' and it is in this atmosphere of doubt, uncertainty and therefore inevitable guesswork that the estimator must work and any 'estimate' prepared by him is necessarily a matter of opinion, arbitrary and open to criticism (especially *after* the building work has been carried out!).

In order to be able to carry out his task effectively, a great deal is demanded of the estimator since the whole of the tendering and estimating process requires him to possess a thorough working knowledge of many aspects of the construction industry, especially with regard to their effects on cost. Those subjects of particular importance are identified below and it is essential for students to appreciate that they cannot be looked at in isolation, but must form part of an integrated study in order to provide the basis of efficient and accurate estimating and tendering.

1 Building technology and services.
2 Management theory and techniques with particular regard to precontract planning, tendering policy and the organisation of resources.
3 Quantity surveying, including an understanding of contract documentation and forms of contract.
4 Economic studies, particularly supply and demand analysis, markets and the effect Government economic policy can have on the construction industry.

The route to the obtaining of a position as an estimator has traditionally been extremely varied. A trade background, for example, will equip him with a 'grass roots' understanding of the practical problems and time needed for building operations on site, whilst a schooling in planning and management would provide an all-round appreciation of the organisational problems involved in the building process, i.e. the preparation for and provision of all necessary resources in the right place at the right time. Yet another route is by way of training and experience as a quantity surveyor. The building up of prices is an inherent part of the quantity surveyor's normal post-contract duties which, together with his knowledge of the cost implications of design, building techniques, materials and contract documentation, make him ideally suited to the job of estimating.

Regardless of the estimator's background, however, the most valuable weapon in his armoury is that of '*experience*' which, in common with any other occupation, can only be gained over the years by learning his 'trade' from other members of the building team and through the correction of inevitable mistakes and errors of judgement made on earlier contracts (though it is hoped that the consequences of such errors are not *too* serious!). Whatever the initial qualifications held by the estimator there can never be any substitute for experience.

His job

In broad terms, the estimator's job is quite straightforward – his principal aim is simply to produce an estimate of the net cost to his company of carrying out a specified quantity of building work within a given period of time. However, whilst the job description may be easy enough to define, anyone thinking that such a task is either 'straightforward' or 'simple' should think again.

The main purpose of this book is to examine the many and varied stages of the building process at which costs are accumulated, including both pre- and post-contract phases, and attempts to provide an insight into the pitfalls and difficulties involved in identifying and calculating the extent of such costs.

For the estimator in practice, the nature of his work is complicated further by a crucial need common to most industries – not only must he be able to locate and quantify costs, but in addition and perhaps more importantly he must at the same time be competitive if his firm is to stay in business. Competitiveness is a requirement which on most occasions runs counter to his desire to ensure that *all* likely costs are adequately covered in his tender.

Therefore the success of the estimating function, whilst initially being associated with the acquisition of work by virtue of producing the lowest tender, must eventually be measured against the ability of the contractor to build within the estimate of net cost and complete fully his contractual obligations emerging at the end of the contract with something like his budgeted profit margin. Unfortunately, of course, there is no way of knowing how accurate forecasts of costs are until those costs are actually incurred, by which time it is too late to take any corrective measures!

Perhaps in the future some new and efficient method of obtaining work may be devised which is universally acceptable to both clients and contractors alike and which will guarantee the selection of the right contractor, remove any uncertainty with respect to client and contractor's costs whilst at the same time giving the contractor a guaranteed return . . . but until that time arrives, the dictates of the traditional system of tendering for work will remain, ensuring the continued existence of, and demand for, the skill and flair of the contractor's estimator.

Tendering and estimating

The two words 'tendering' and 'estimating' are often used in the industry to mean the same thing; however, a distinction should be made since the **Estimate** is the preliminary assessment of the net cost of carrying out a specified amount of building work whereas the **Tender** is the final price or 'offer' which is submitted to the client by the contractor and is the sum of money for which he is prepared to carry out the work and will include not only the 'estimate' but also a margin for overheads and profit.

Whilst the preparation of the estimate is largely a relatively mechanical operation involving the calculation of labour, material, plant and other site operating costs, the conversion into a tender by the addition of a profit margin is a delicate task often undertaken by management at an 'adjudication' meeting where aspects other than pure cost are considered and a decision made on the extent of the mark-up to produce a tender which must be both competitive and profitable for the contractor.

Chapter 1

Contractor selection

One of the most important tasks facing a client when embarking on a scheme of construction work must involve making the correct choice of contractor for the project and to enable him to do this a number of different methods are available to suit a variety of circumstances, the most common system of contractor selection utilising the traditional bill of quantitics which together with a number of corresponding drawings can be used to generate competition between rival contractors in order to obtain keen prices. However, whilst this method is perhaps the most common, there are many instances of where such a system cannot be used nor would it be in the client's best interests in all cases to follow the traditional path even if it were feasible. The correct choice of the method employed in selecting a suitable contractor and the contractual arrangements to be adopted is of vital importance if the client is to obtain the best value for money in the light of the particular circumstances surrounding his project. Each contract is unique and is worthy of careful individual consideration with the 'traditional' method chosen only where an alternative system cannot be seen to give the client a better deal, and to this end heavy reliance will be placed upon the expert professional advice of the architect and quantity surveyor.

Irrespective of the final contractual arrangements chosen by the client, the method of selecting the contractor must first be established, the two basic methods being:

1 By Competition
2 By Negotiation

1.1 Competitive tendering

A competitive tender refers to the final price submitted by the main contractor which itself may be made up of both competitive prices, i.e. the estimator's own bill rates and negotiated prices, i.e. that portion of the work which may be sub-let to specialist 'domestic' sub-contractors some of whose rates may be determined by negotiation. Most contractors' tenders contain an element of both; however 'competition' in this instance refers to the main contractors' tender only. Acknowledging the fact that negotiation is available to the client as an alternative method of contractor selection, prices obtained from a number of contractors, all competing genuinely with each other for the work, will probably give the best value for money and will be suitable for the large majority of building projects. Although it is generally accepted that the basis of competition is normally related to the process of obtaining the lowest possible price for a job, this need not necessarily always be the case, even though the Code of Procedure for single-stage selective tendering argues for the element of competition being limited to price only with all other possible variables being determined beforehand and stated clearly in the tender documents. Such a recommendation assumes that the client regards cost as the all-important factor in the building process and whereas it must be recognised that in most cases this would be so, occasionally a client may be faced with circumstances which throw a different light on the problem resulting in other factors being made more crucial than cost to the eventual success of a development. For example, speed of erection may be the critical requirement where, say, a manufacturer needs 10,000 square metres of storage space urgently to house a cancelled export order or overproduced stock in order to avoid large-scale wastage. In such a case cost, and perhaps even design and choice of materials, may take second place to the need to erect a warehouse with the specified floor area in as short a time as possible, and so providing tenders are within an acceptable range to the company, the contractor quoting the earliest possible starting and completion dates may well be awarded the contract. Time, being the controlling factor in this instance, would be the basis of the competition. Other forms of competition, for example, the quoting of a percentage increase on top of a schedule of basic prices (term contracts), are considered later.

Having decided to award the contract on a competitive basis, the client must further consider whether to choose 'open' or 'selective' tendering as a means of obtaining his tenders, each method having both its advantages and disadvantages.

1.1.1 Open tendering

The system of open tendering, favoured by many large public bodies such as local authorities, health and water authorities etc., provides for any contractor who feels he is capable of performing the work to submit a price. In order to attract a response from prospective tenderers, the client advertises in the local press, or if the job is a sizeable one, in the national press and the various building journals giving a brief description of the proposed works, anticipated starting date and sometimes an approximate value of the job. (Currently, in accordance with an EEC directive, all public works contracts exceeding approximately £650,000 must in addition be advertised within the member countries of the Community.)

Thus the notification of the proposed works is brought to the attention of any builder large or small, good, bad or indifferent, local or national, who cares to search through the appropriate publications. In this way, by allowing *all* interested parties to tender, the client doubtless can gain the biggest possible response to his advertisement thereby achieving maximum possible competition and almost certainly the lowest possible price. Whilst this may delight the client who initially at least foresees a considerable saving on his development budget, such optimism may well be short-lived with eagerly anticipated savings failing to manifest themselves in the final account when and if it is ever produced! By accepting the lowest tender regardless of any other considerations, the client can often find himself entering into a contract with a local builder and his handful of workmen who, having been in business for all of six months and now possessing an ageing pick-up truck, second-hand concrete mixer, two or three ladders and a wheelbarrow considers himself fully capable of successfully completing a multi-million pound health centre within twelve months. Whilst applauding his ambition and optimism, the most likely outcome of such a venture would be the contract still being in progress five years later or worse, bankruptcy for the builder within a matter of months resulting in the contract having to be re-let at a substantially higher cost to the client once the financial tangle of a bankruptcy situation had been sorted out. Whilst such an example may be an extreme case, the risk nevertheless still remains.

Open tendering therefore leaves much to be desired when from a tender list of perhaps 40, 50 or even more would-be contractors in difficult times, the client has to look to perhaps the sixth lowest to find the name of a reputable builder who could be entrusted with the job! It is very tempting for a client to accept the lowest price irrespective of the reputation or suitability of the contractor who may not be properly equipped to undertake the work resulting in a poor quality building. One further point which is probably overlooked by the client is the large waste of tendering effort with too many firms pricing. On many occasions, the cost of preparing a tender could amount to several hundred pounds and where specialist engineering contracts are involved where a large degree of design work must be done by the tenderers, a four-figure sum may not be uncommon. Thus where an open invitation to tender attracts, say, 40 contractors all of whom submit a bona fide tender (which is highly unlikely) 39 of them will be unsuccessful and where the cost of preparing the tender is say, on average, £500 for each one, the cost of abortive tendering on that one project will amount to nearly £20,000. In order for each contractor to stay in business, the cost of preparing unsuccessful tenders must form part of that company's overheads and subsequently be incorporated into future, successful tenders. Obviously the greater the proportion of unsuccessful tenders to successful ones, the higher will be the level of tendering generally. Open tendering with its large tender lists and significantly longer odds of being successful contributes to the higher level of waste and therefore the cost of building in the country as a whole. In an attempt to reduce the tender lists, clients often ask for a deposit to be lodged which is then returned to the contractor on receipt of a bona fide tender. This deters contractors not seriously interested in pricing the work from applying. The amount of the deposit, however, must be high enough to deter such people and yet not be so high as to discourage a genuine tenderer!

1.1.2 Selective tendering

Many of the vagaries of the open tendering system can be overcome by adopting a method known as 'selective tendering'. Competition between contractors is still maintained by the client whilst at the same time unsuitable builders are eliminated from the process. Again, as with open tendering, this system is not without its faults since by restricting the tender list to perhaps local firms with a good reputation who are capable of complying with the contractual obligations, the level of competition is necessarily reduced. Furthermore, such firms are the ones which will carry a good stock of plant and equipment, employing experienced technical and managerial staff, and can provide sufficient back-up resources to ensure the satisfactory completion of the contract. The overheads of these firms will tend to be high and a reasonable profit margin will be expected and will thus possibly be reflected in higher prices. In return, the client is guaranteed a good standard of work, with the probability, all other things being equal, of the job being finished on time. If the contractors are carefully selected in the first instance, the extra cost involved would be far outweighed by the benefits to the client. If this is accepted, the main and most important task then may seem to lie in preparing the list of contractors who are to tender, and if this is done well, the client should be happy for any one of them to perform the work and in this case the lowest price *only* will be the deciding factor. How then is this 'selected list' compiled? Firstly a list of firms who are suitable and capable of undertaking the contract is prepared, preferably with the architect, quantity surveyor and client all in consultation with each other – the client may insist on a particular firm or firms being included on the list. A private client is fully entitled to do this. Alternatively, an advertisement may be placed in the press, inviting *any* contractor, as before, to apply for the opportunity of tendering. However, this system differs from open tendering in that from this initial list, only certain competent contractors will be invited to submit a tender. The Code of Procedure for single-stage selective tendering suggests the following scales in determining the optimum number of tenderers for varying contract values:

Size of contract	Maximum number of tenderers
Up to £50,000	5
£50,000 to £250,000	6
£250,000 to £1 million	8
£1 million upwards	6

Note: These figures were determined when the code was first published in 1977 and should be adjusted accordingly for inflation.

Such a scale would require revision from time to time in order to take inflation into account. When selecting the contractors for the short list, the following items should be taken into consideration:

1 The firm's financial standing and record.
2 The firm's recent experience of completing similar work within the specified time.
3 The structure of the firm: technical and managerial staff, workforce and back-up facilities.
4 The firm's capability of accepting the work at the required time.

Where a client maintains a selected list of contractors as opposed to the preparation of one on an *ad hoc* basis (i.e. a 'one-off' job), such a list should be reviewed from time to time to exclude those companies whose performance has been unsatisfactory and to allow the introduction of new suitable firms who would like to be given the chance to prove themselves capable.

1.2. Negotiated tendering

Having stated that competition (preferably by selective tendering) is acceptable under normal circumstances, there are occasions when it would suit the client to dispense with the element of competition altogether and approach only one contractor who is considered to be the most suitable under the circumstances and negotiate a price with him. Some circumstances under which negotiation may be preferable would be as follows:

1. A quick start required

Taking the example of the manufacturer earlier, the need for an early start to the building of the warehouse may be so pressing as to warrant the saving of the time needed to go out to competitive tender. A contractor can be selected almost immediately and the work can begin whilst the final design and cost are still being formulated. Again, from an earlier example, the dubious builder, having been selected by open tender and finding the going tough electing to go into liquidation could pose a serious problem for the client who now finds himself in financial difficulties needing continuity of work if he is to avoid further losses, since he will be unable to gain an income from the building until it is completed and fulfilling its planned function. The extra cost which would doubtless result from another contractor stepping in at short notice to finish off someone else's work, with all its attendant problems, may well be offset by the saving in time.

2. Business relationship

There may exist between two companies, not necessarily working in the same field, a good relationship of long standing where a measure of reciprocal trading benefits both parties. For example, a contractor may agree to buy all of his company cars from a particular car distributor on the understanding that should the dealer require further office accommodation, warehouse facilities or repair and maintenance work, the contractor would get the business.

There may be a joint partnership in the development of, say, a block of luxury flats on a plot of land where a finance company is prepared to finance the venture and the contractor owns the land. In such a case it is likely that an independent quantity surveyor may be appointed to establish a fair price for the work or monitor the costs during construction.

3. Continuation contract

Under circumstances where a client envisages a new project, very similar in size and design to one recently completed and again, if a good relationship existed between the client and the contractor, the latter may be invited to negotiate a price for the new building. He may well be able to effect savings since being familiar with the design and contruction, he will be aware of the difficulties previously encountered and be able to use his knowledge of the work to improve efficiency on the new job. He may well have constructed some special formwork which could be used again, saving costs and perhaps improving on the completion date.

4. The state of the market

When the building industry is overstretched and contractors are kept busy with full order books, a client may decide to opt for the lesser of two evils and choose a contractor with whom he can negotiate, rather than paying a heavy premium in attempting to persuade other contractors to price who are not really interested or who haven't got the spare capacity, even if successful by competitive methods to take on the job. The choice of a suitable contractor would depend upon the type and size of the job in question. In remote areas there may be only one contractor available if the job is not sufficiently attractive to interest outside contractors. In this case a client would have little choice but to accept the situation and negotiate with the local company.

5. *A company having specialist plant or techniques*
Where a large organisation has developed and patented a special technique, design or piece of equipment which is considered to be a market leader and can be used to good effect in the building process, whether to increase the speed of erection or to work more efficiently or cheaply, a client may find himself better off by resorting to negotiation with the contractor and taking advantage of his expertise rather than by using more traditional methods. A good example of this is the familiar 'no-fines' system of constructing concrete walls patented by a large well-known building contractor which enables the concrete shell of a house or block of houses to be constructed very quickly and cheaply. Such a system could be taken advantage of by, say, a housing association, or local authority who, working on very tight budgets, wish to provide as many housing units as possible for their limited amount of money.

The above examples give an indication of when it may prove beneficial to a client to negotiate rather than choose competitive tendering. However, the list is by no means exhaustive; nor should it be assumed that negotiation would be preferable in all cases. Each proposed development should be looked at individually and a conscious decision made in the light of each set of circumstances bearing in mind the criteria which is – 'benefit to the client'.

1.2.1 Techniques in negotiating

Having decided to award a contract by negotiation, how can this be satisfactorily achieved without presenting the 'fortunate' contractor with a blank cheque? Indeed, where a patented system of building is chosen intially because of its critical contribution to the project, this may to some extent be unavoidable where the contractor concerned is the only one in the field or where there is very limited competition. (Such a situation often occurs in work involving specialist engineering services such as refrigeration pipework and plant, lift installation etc.) However, leaving aside the situation where the prospective client is presented with a *fait accompli* of list prices and 'take it or leave it', before normal negotiations can commence a 'starting point' must be agreed upon. Basically, one of two methods may be chosen, namely:

1 A priced bill of quantities for a similar project previously carried out under similar conditions is adopted which serves as a 'foundation' for determining the price of the new building. The bill of quantities adopted in this way is known as a *nominated bill*.
2 By attempting to predict the final costs of construction at the outset and adding to those costs an element for overheads and profit.

1. The use of a 'nominated bill'

The use of this method as a basis for negotiation, though seemingly straightforward, is not without its difficulties, perhaps the main one being the choice of a 'suitable' bill of quantities favoured by both parties, a problem which in itself could result in protracted negotiations! Obviously what may be considered to be a suitable bill by the contractor may prove to be totally unacceptable to the client and both sides need to study carefully not only the rates contained in a possible bill but also the conditions under which it was priced.

Taking such difficulties into consideration, however, if both sides adopt a reasonable attitude with a little 'give' and 'take' on both sides (which after all is the means by which most building disputes and differences are solved!) adjustments can be made to a likely bill of quantities in order to allow for discrepancies which will inevitably occur between the chosen 'similar' job previously carried out and the proposed project.

For example, were the contract conditions particularly onerous or was the competition unusually fierce? There may have been local difficulties at the time the job was priced such as shortage of labour, a steel strike, shortage of bricks or timber, all of which may have produced anomalies in the pricing of certain items.

It is unlikely that the priced bill of quantities to be adopted will contain *all* the items of work which will be encountered in the new project and where this occurs new rates will have to be built up from scratch and agreed between the two parties. This should not cause too many problems providing the differences between the two jobs are not *too* great and the *bulk* of the new work can be valued using the adopted bill rates suitably adjusted for inflation.

Allowing for the fact that the job is negotiated, which automatically implies a higher cost to the client, at least by using a nominated bill in this way the client can be satisfied that the rates used to evaluate his job are based on prices originally obtained on a competitive basis and should therefore reflect a fairly conservative method of evaluation and provide the best system for accountability.

2. Assessment of costs

Not all new projects will lend themselves suitable for evaluation by the use of a nominated bill. For a 'one-off' project there may be little chance of finding a previously constructed 'similar' job which can be utilised. In such circumstances the client and contractor may agree between themselves what the final cost is likely to be which, when achieved, will provide the contract sum once an agreed element for overheads and profit has been added. The process by which an assessment of the possible costs can be achieved is not far removed from the method which would be adopted by an estimator himself when preparing a tender. Where this method applies, rates, both unit and composite to suit the conditions, are built up from scratch using suppliers' prices for materials and goods, all-in rates for labour and sub-contractors' prices for work which would be sub-let. Again, as with the use of a nominated bill, this method presents its own problems. The process could be lengthy since some quantities must be prepared (however rough!) from a set of drawings (however sketchy or incomplete!) in order for the work to be priced even though the two parties would be working together to produce the result. The establishment of the unit rates which initially will be at net cost will undoubtedly lead to a difference of opinion since at best such rates are only a *prediction* of the *likely* cost to the contractor and are by their nature purely arbitrary, the exaggeration of the risk and problems foreseen by the contractor being equally matched by the client's optimism as to the speed and ease at which the various building operations can be performed! Inevitably such a situation can lead to friction and frustration between the two quantity surveyors and provides all the ingredients for many sessions of hard bargaining; however, it must be remembered that the essence of negotiation is the conferring of both parties with a view to compromise and a 'swings and roundabouts' approach must be adopted if this is to succeed. It is essential that both parties should be able to negotiate from a position of strength and this can only be achieved where the protagonists involved are able to act with the same degree of authority on behalf of their respective employers. For example, negotiations handled on behalf of the client by, say, a senior partner of a quantity surveying firm should be countered by a person at director level representing the contractor. Equally important is the need for both parties to be in full possession of all relevant documents, quotations etc., which will be used as the basis of the unit rate calculations. Since it will be the contractor who, being more familiar with his own costs of production, regular suppliers and so on, will obtain all such information in the first instance, his integrity must be relied upon in ensuring that copies of *all* relevant documents are passed on to the client. Once an attempt is made to conceal, for example, a 40 per cent discount allowable by a supplier on some particular wallpaper chosen by the client, the feeling of trust will be destroyed, which could at worst lead to the breakdown of negotiations.

Whether a nominated bill or an agreed assessment of the costs is chosen as the method for valuing the work, it must be decided before any detailed negotiations can commence just how the final price is to be determined, since many other items besides the unit rates themselves need to be included in the overall price. For example:

(a) Profit and overheads. Some contractors prefer to include these as a lump sum in the preliminaries section of the bill, often split up and disguised under a number of unlikely headings, whilst others prefer the more usual method of including a proportion of profit and overheads within each unit rate thereby distributing this element more or less evenly throughout the whole bill.

(b) Plant and equipment. Again contractors have the option of either including the cost of plant, scaffolding etc., either as a lump sum in the preliminaries or incorporating the cost of such items within the unit rates of appropriate trades which utilise them, although bills of quantities now being prepared under the sixth edition of the *Standard Method of Measurement* provide for the cost of any items of plant etc. to be priced as a lump sum within each separate work section.

(c) Firm price contracts. Where a job is to be priced on a 'firm price' basis, the usual practice is for the estimator to include an element for inflation within each unit rate of the bill. This enables him to be more flexible in the allowances he may make for increased costs within different trades. However, there are no hard and fast rules to be followed and a contractor may well choose to price all of his unit rates at current-day prices and allow an overall lump sum, again probably somewhere in the preliminaries bill, in order to cover all likely increases in cost.

(d) Tender adjustments etc. Any tender adjustments a contractor may wish to make in respect of risk element, eleventh hour changes in pricing policy, working methods etc., are usually covered by a lump sum, calculated and inserted somewhere within the bill, perhaps once more in the preliminaries.

Assuming a nominated bill is chosen as the basis for negotiations, there must be contained within that bill areas of cost, of which the above four items are good examples, which are not immediately apparent without further investigation. The bill must therefore be broken down to determine the basis upon which the original tender was formulated. Once this 'structure' has been determined and agreed upon the negotiations can commence with the greatly reduced risk of arguments occurring later on when variation orders may require valuing. Similarly, if the agreed assessment of cost method is to be used, once the structure of the proposed estimate has been agreed, rates can be built up from scratch with no difficulty.

Chapter 2

Contractual arrangements and documentation

Having considered the methods which a client may use to select his contractor, i.e. by competition or negotiation, there remains the decision as to which form of tendering is to be used, which contract to use as a basis for the 'building agreement' and what system of contract administration is best suited to the particular set of circumstances surrounding the project. The options open to the client which are discussed in detail below include the most common alternatives available.

2.1 Traditional method

The 'traditional' method of tendering with which the industry is most familiar is perhaps the most convenient and satisfactory providing a number of preconditions have been fulfilled. The traditional method involves a bill of quantities, fully detailed and prepared in accordance with the latest edition of the *Standard Method of Measurement*, being supplied to each one of a number of contractors who have been (preferably) selected by the client to price the work in competition with each other. For such a system to work successfully it presupposes that for an accurate bill of quantities to have been prepared there must necessarily have been a great deal of finalised design work completed by the architect, which in turn depends upon the client having clear ideas of exactly what he wants at an early stage in the operations. Furthermore, in addition to the design work being completed and bill of quantities prepared, the use of this method also assumes a final choice of form of contract and contract conditions (including firm or fluctuating price, starting date, completion date, liquidated damages, etc.) has been made together with any other obligations or restrictions the client wishes to impose upon the contractor. Thus, the success of the traditional method is only possible where major preliminary design work has been achieved. Where the drawings and bill are incomplete, vague or ambiguous or where the conditions of contract have been determined with little thought as to the consequences, the use of the traditional method will prove disastrous and no doubt the contractor appointed to carry out the work will have a field day reminding the client, to his cost, that the tender figure was based solely upon the information presented to him at the time of tender. The traditional method is also the most costly for the client initially, since architects' and quantity surveyors' fees must be allowed for and it takes a great deal of time for the drawings, bill and specifications to be prepared accurately. The time allowed for tendering (say three to four weeks) must also be added to the time for the preparation of the tender documents, and where time is of the essence to the client, the need to make an early start on site may well preclude the use of this method. In situations where the foregoing conditions do not present a problem to the client, the orthodox method can be relied upon to display a number of important advantages, namely:

1 Since the tender documents are prepared in advance and each contractor receives identical information at tender stage, then each of them will be pricing on the same basis, i.e. using the same descriptions and the same quantities thereby avoiding artificially low prices from a contractor who may have made different assumptions, or prepared his own quantities, which subsequently prove to be inaccurate.
2 Where work is to be carried out in phases with sectional completion of various parts of the project required at different dates throughout the contract period, the additional costs relating to this requirement can be established at an early stage by filling in the relevant details of the appendix to the conditions of contract (assuming the JCT Standard Form of Contract is used) and thereby bringing this contractual obligation to the contractor's attention at tender stage. On the submission of tenders, it can then be assumed that all the contractors have considered the point and included a sum within their prices to allow for this. The advantage of making a condition, such as phased work, clear to contractors at tender stage is that there is a good chance, given that the contractors in competition with each other are keen to obtain the work, that they may tend to reduce or even eliminate the risk entirely in the hope of keeping one

step ahead of the opposition. A client who decides he would like phased completion once the contract has got under way or even once tenders have been received is likely to find himself with a major claim on his hands.

3 Once the successful contractor has been chosen and his priced bill submitted and accepted, at least the rates contained within the bill will, wherever possible, be used to price variations thus providing a fairly accurate measure of cost control and final account prediction.

4 In the absence of major variations, the orthodox method of tendering will present the client with a tender figure which should not differ greatly from the final account since all or most of the work is detailed and priced at tender stage – an important consideration where the client's budget is restricted.

2.2 Bills of Approximate Quantities

Where detailed drawings are unavailable at tender stage, accurate bills of quantities obviously cannot be produced, but nevertheless a bill of *approximate* quantities may be prepared where the scope and scale of the work are fairly well established at this point and a design and specification for the works and materials have been produced. Contractors will submit their tenders based on the approximate quantities with the actual being remeasured from detailed drawings as they are produced during the course of the contract or physically measured on site with payment being made accordingly. Although the initial tender figure will be superseded by the value of the remeasured works later on, bills of approximate quantities do, at least, give the client a fair estimate of the likely final cost, and where the actual work does not deviate too much from preliminary plans the contractor is given a sound basis upon which to gauge the extent of his requirements for the project with rates inserted against items in the bill providing a realistic basis for measuring and valuing the subsequent building work carried out. However, under circumstances where large variations to the anticipated work are likely, the contractor may well be justified on many occasions in revising his prices accordingly, with the result that the final figure may be far removed from the initial tender sum. This system is useful where a client wishes to make an early start on the construction without waiting for full details and bills to be prepared.

2.3 Cost reimbursement contracts

On occasions it may not be possible to ascertain by measurement the full extent and nature of the work involved before the contract gets under way, the design work remaining incomplete at tender stage.

However, the client may still wish to appoint a contractor and make a start on the job as soon as possible resulting in decisions and final design work being determined as the contract progresses. In such circumstances it is impossible therefore to use the traditional method of contractor selection, there being no detailed drawings or bills of quantities from which a firm tender figure can be obtained in advance. One option open to the client is a method of reimbursing the contractor on a suitable 'cost plus' basis where the actual (rather than the predicted) costs of construction are paid, together with a 'fee' which represents the contractor's management and technical costs and overheads and profit. The features of this system are as follows:

1 It encourages a greater degree of co-operation between contractor and client with both parties perhaps working together closely in the design stages and decision-making as the project develops, the client being able to take advantage of the contractor's knowledge and expertise. With both sides working together in this way, fewer disputes can be expected.
2 Since the contractor is paid in accordance with his actual costs incurred in building the project, it is essential that an efficient system of recording such costs (known as 'prime costs') be devised whereby both the contractor and client are satisfied that 'fair play' is being achieved. To this end, weekly time sheets verified by the architect or clerk of works must be submitted to establish labour costs, quotations, delivery tickets and invoices for material costs, and plant returns for machinery and equipment used carefully kept. Although an accurate record of the mounting costs can be maintained week by week, the paperwork involved in this procedure can be very costly with both client and contractor having representatives based full time on site engaged in this task.
3 The contract administration in general presents a problem and close co-ordination must be maintained between the architect and the contractor in establishing the optimum level of manpower and the extent to which plant and equipment is needed at each stage in the development. If this objective is not achieved, the inefficient contractor will be rewarded since there is little incentive for him to keep his costs down, particularly where the fee is to be calculated as a percentage of final costs!
4 A cost reimbursement contract need not necessarily be entered into on a negotiated basis; an element of competition can be introduced into the method of choosing the contractor by allowing a number of selected contractors to quote their own fee which can be either a fixed lump sum or expressed as a percentage of total costs. Either way, the contractor quoting the

lowest sum or percentage would be awarded the contract. However, in order for competing contractors to arrive at a realistic fee, an estimate of the likely extent of the prime costs involved must be prepared by the client's quantity surveyor and must form part of the tender documents although in the event such an estimate can be only very approximate with design work and other information being incomplete at this stage. Despite this drawback, it is important that the original estimate is assessed as accurately as possible in order to achieve some measure of cost control over the job. At regular stages thoughout the contract, the value of the estimated cost will be compared with the actual cost to date, with the difference between the two figures indicating the extent to which the budget is overspent or the amount of savings made. Having this information to hand at any given time will enable the anticipated final account figure to be constantly updated which when relayed to the client may well influence his decision-making and level of expenditure on the remainder of the project.

5 The valuing of variations presents no problems as the costs involved in design changes etc. are simply absorbed into the total figure as the work proceeds, although the cost of rectifying defective work must be clearly identified and kept out of the reckoning!

6 The contract may be executed not only by the main contractor himself, but also by nominated and private sub-contractors whose work may be valued on a different basis necessitating the additional task of separating such work and monitoring their costs as a separate operation.

The main drawback of a straightforward cost reimbursement contract is the lack of incentive for a contractor to minimise his cost; however, this can be overcome by developing the contractual arrangements a step further to produce a target cost contract.

2.4 Target cost contracts

The method employed on a target cost contract is similar to that used for the cost reimbursement type in that a provisional estimate of the prime cost is prepared, but this time the figure is agreed as being realistic by both the contractor and client before work commences. To this sum is added the contractor's fee for overheads and profit as before and the resulting figure is adopted as the 'target cost' for the work. However, an opportunity for an added 'bonus' to be earned by the contractor is introduced since reimbursement is made only *partly* on actual costs with a final adjustment being made in relation to the amount by which the actual cost, including the original fee, either falls short of or exceeds the target figure. In practice, this could result in the contractor receiving an extra payment over and above his original fee in the former case or being penalised and having to shoulder some of the extra cost in the latter. Looking at the positive side, this system could appeal to both parties since from the contractor's point of view, the lower the actual costs, the higher his bonus will be, and for the client, despite having to pay out a bonus to the contractor, the final account will still fall short of his budgeted figure, representing to him a saving on his anticipated expenditure. Unfortunately, however, not all contracts run according to plan with site problems, weather or contractor's inefficiency all serving to push the final cost beyond the target figure. In this case both the contractor and the client will lose out since, again, final reimbursement is determined only partly by *actual* costs.

The extent of the actual saving or additional cost, as the case may be, to either party is mitigated to some extent by the effect of a predetermined 'share-out' of the money involved. For example, any savings made would probably be due in part to the architect's final design and in part to the contractor's good management and efficiency, in which case savings may be shared equally between client and contractor; similarly, any loss incurred being established on the same basis. In cases where the contractor is called upon to accept a greater degree of risk perhaps involving an element of danger as well as uncertainty, an equitable split may be 70–30 per cent in his favour. Alternatively, the actual percentage can be introduced as a further competitive element into the tender in addition to the estimation of the prime cost, although in these circumstances it would be difficult, if not impossible, to decide which of two contractors would eventually be the cheaper where the estimated prime cost of one is lower but his required percentage of any savings achieved is higher! The target cost contract certainly provides a much greater incentive to the contractor to work efficiently and reduce his costs; but at the same time, in trying to achieve this objective, careful supervision is required to ensure that the quality of the work does not suffer. Having agreed the amount of the target cost with the contractor, the contract administration work is the same as for the fixed fee cost reimbursement type with a careful and accurate record of all costs being kept as the contract progresses. The greatest difficulty, however, is in obtaining a fair and reasonable estimate of the prime costs since contractors will naturally attempt to inflate the figure as much as possible thereby gaining the maximum advantage where large-scale savings are subsequently achieved. The client therefore needs expert advice from skilled and experienced quarters to ensure that prices whether obtained by negotiation or competition are realistic.

A variation of the normal target cost contract can be used where early completion is essential, the contract period being the critical ingredient of the project with cost being a secondary consideration. In this case, 'time' would be the subject of the target with contractors competing by quoting anticipated completion dates, and bonuses or penalties based on the amount by which actual completion of the work falls short of or exceeds this date.

2.5 Continuity contracts

On occasions where a client envisages further work beyond the immediate project, he may wish to maintain continuity of work with the same contractor. An ideal opportunity to make use of this sort of arrangement would present itself where a long-term large-scale housing scheme is planned involving, say, 300 dwellings. Such a large contract if attempted as a single job would not only be extremely costly, but also difficult to manage with any one contractor having to commit a great deal of resources to that job. An alternative approach would be to award the work piecemeal to three or four contractors, each being comfortably able to carry out his allocation of work without overstretching himself, although this of course assumes that the client has the funds available to tackle the whole of the work immediately. Where this is not possible, the scheme is likely to be split up into say three phases, each one comprising 100 dwellings enabling the client, if he wished to treat them as three separate and independent contracts, obtaining orthodox competitive tenders for each and possibly ending up with three separate contractors. However, the advantages to be gained by selecting the contractor for the first phase only by normal means and thereafter negotiating with that firm for the remaining two phases as and when they can be started can be summarised as follows:

1 Contractors would be informed when pricing for Phase 1 on a competitive basis that further work would be awarded to the successful tenderer providing the work is of the required standard and is completed within a reasonable time. Promise of continuity of work for perhaps two or three years ahead would be a very attractive proposition to contractors and a keen tender list would be anticipated with firms perhaps pricing at little more than net cost being anxious to secure not only the immediate contract but also the future work in the pipeline. Obviously where future prices for the remaining phases are negotiated using low prices as a starting point, the client can be satisfied that he is obtaining good value for money, although where the original price is *too* low, an otherwise good relationship between contractor and client could be damaged where it becomes impossible to make a profit, despite the advantage of negotiation.

2 The security of continued production which the contractor enjoys under this system will encourage a good working relationship between himself and the client.
3 There will be a saving to the client in subsequent tender preparation costs since the original bill of quantities can be used where the future work is similar, using the prices contained in the bill as a basis for negotiating the value of the remaining phases.
4 Despite having to negotiate prices for the subsequent phases, the client may still expect a saving compared with formal competitive methods since the incumbent contractor should be able to carry out the second and third phases much more efficiently and quicker than a newcomer, being now familiar with the design and having learned from problems previously encountered. There may be savings on some preliminary items which can be utilised on future phases such as the site huts, compound, telephone etc., also items of formwork specially fabricated for the original contract could be re-used. The extent of any savings thus achieved may be difficult to quantify, but they should at least serve as a useful tool to aid the client in his negotiations! The continuity contract is ideal for a 'one-off' situation where the likelihood of future work exists, but where no guarantees can be given, if the client is subsequently unable or unwilling to proceed with further plans.

2.6 Serial contracts

A serial contract differs from a continuity contract in that at tender stage, rival contractors are informed that the successful firm will be called upon to carry out a number of future separate contracts each of which will be very similar if not identical to the original one being priced, the prices contained in the bills of quantities being used to value similar work on the future projects. The approximate extent of the 'series' of contracts will be known at the initial tender stage even though the design work on some of them may as yet be incomplete. Again the advantages to be gained from maintaining continuity of work with the same contractor are the same as outlined above, but serial tendering being in effect a standing offer to carry out a series of projects all based on the priced bills of quantities for the first project which becomes the 'master' priced document. Rates contained in the original bill of quantities can be updated to account for inflation using either a 'conventional' or 'formula' method for assessing the amount of increased costs. The guaranteed workload to be enjoyed by the successful contractor should ensure that the client obtains keen tenders from competing contractors in the first instance.

2.7 Term contracts

A term contract is a type of continuity contract which has been adapted to suit situations where a continuous programme of work is required on a particular site or within a certain defined area. This system is particularly applicable to Government establishments, bases for the armed forces, large-scale industrial sites and oil refineries etc., where there will be a constant need for maintenance and repairs together with small and medium-scale extension and alteration work. When tendering, contractors are aware of the nature and approximate value of the work based on previous contracts, but at this stage the actual extent of the work cannot be known and tenders are submitted in the knowledge that the successful contractor will undertake to carry out all the work given to him over a certain period of time or 'term', a typical period being perhaps two years. The tendering document will normally be a schedule of rates covering most of the work normally encountered with contractors quoting a single percentage addition or reduction on the schedule as a whole to cover preliminaries, overheads and profit. The successful contractor will be the one quoting the lowest percentage and orders will be issued to him from time to time by the supervising officer or clerk of works and the work valued in accordance with the schedule rates amended by the percentage addition/reduction, each works order having a value of between say £10 and £30,000. On expiry of the two-year term, the contract will be up for renewal and the contractor selection process adopted once more; however, it is not unusual for one contractor to secure the work time after time, having been established on site for perhaps twenty years or more and knowing from past experience the right level of percentage addition which will ensure a profit, it being extremely difficult for an 'outsider' to judge this with any degree of accuracy.

2.8 Two-stage tendering

As the name suggests, prospective contractors take part in a selection procedure involving a process of elimination before the most suitable one for the job in hand is chosen. Such a system may be used by a client planning a complicated project where the type and scope of the work only is known at this early stage and where he would benefit from the advice and expert knowledge of a contractor experienced in that particular type of work and who could help in the design and planning of the contract. A number of possible contenders would be selected initially and each would have the opportunity at pre-tender meetings to discuss with the client's professional advisers their individual ideas regarding approach, methods of tackling the work, programme, management and organisation etc. Having satisfied himself as to the suitability of some or all of the tenderers, the client can then introduce a further element of competition into the selection procedure by inviting them to price a schedule containing approximate quantities of some of the more common operations which will form a major part of the work. If a bill of approximate quantities can be prepared for this task, so much the better. On the evidence of the preliminary pricing exercise, the lowest tenderer will then be invited to join the team planning the project and be expected to take an active part in giving advice on cost comparisons, programming and site organisation and generally assisting in the design which when finalised will enable accurate bills of quantities to be prepared and a value for the work determined based on the prices contained in the original schedule. Care must be taken to ensure that the level of pricing in the final bill of quantities is consistent with that in the schedule, otherwise the final tender figure could be distorted out of all proportion. In order to achieve this, it is essential for the client's advisers to have a thorough knowledge of builders' estimating techniques and for the contractor to share his information willingly and where necessary showing how individual prices are built up indicating the location of overheads and profit. However, set against this drawback is the likelihood of the contract running smoothly and finishing on time, with very few subsequent variations to impede progress. With regard to the contractor's contribution during the design stage, reimbursement for services rendered here could be made on a separate basis.

2.9 Design and build (package deal)

With this type of arrangement, a client invites a number of selected contractors not only to price the work, but also to design, plan and organise the contract. The successful contractor would then provide an 'all-in' service or 'package deal' from sketch schemes, artists' impressions through the construction stage to handover, being fully responsible for all contract administration and planning. The benefit to be gained from this approach is that by carefully choosing contractors who are especially experienced in the type of work envisaged, advantage can be taken of their expertise by allowing them the flexibility to produce their own schemes. In this way, in the early stages at least, the client need not necessarily be restricted to negotiation, but can encourage competition not only on the basis of price but also on the aesthetic qualities of the schemes put forward, contract period offered, costs in use etc., with each contractor trying to 'sell' his own individual 'product'. Naturally, the client will pay for the contractor's design and planning costs which will be included in the lump sum 'all-in' price; however, a contractor well versed in large-scale traditional housing or industrial warehouse units can often produce a more cost-effective scheme than a client's own architects. This method of contractor selection should suit a client whose main concern is that certain minimum standards, in accordance with his own performance specification, are met by the contractor. However, great care must be taken when preparing such a brief, and professional advice should be sought regarding not only its content but also to ensure that the tenderer's own schemes meet the requirements. The performance specification, whilst still giving contractors a relatively free hand in the design, may centre around the construction of, say, a given number of housing units each having a certain floor area, number of rooms and possessing all the usual amenities described in detail or it may relate to a warehouse where the provision of a specified storage area sufficient to carry a given loading is the central issue. Once appointed, the successful contractor can be expected to make good progress working to his own well-tried and familiar design, thus virtually eliminating variations, delays and disputes with the work finishing on or before time.

2.10 Drawings and specification

For small works involving mainly extensions and alterations, contractors are often asked to submit a price on the basis of one or maybe two drawings either with or without an accompanying specification. Tendering under this method presents contractors with additional problems and risks related mainly to lack of information both in a technical and contractual sense. Drawings prepared for a client are often undertaken by inexperienced draughtsmen who do not fully appreciate the kind of information needed by a contractor for the purpose of preparing an accurate and realistic price. Builders subsequently pricing on the strength of inadequate or incomplete drawings will have to try and interpret the client's requirements as best he can with the result that each contractor is often pricing on a different basis, particularly if items are missed or wrongly measured. In order to protect himself legally, the estimator would be advised to write his own detailed specification, including any assumptions made and stating clearly the terms upon which the tender is based such as fixed or fluctuating price, starting and completion dates, methods of payment, retention etc. The two-fold exercise of taking off quantities and pricing them adds greatly to the cost of tendering and so the estimator will have to try and gauge the genuineness of the enquiry. On occasions, would-be clients are simply interested in obtaining a budget figure with little or no real intention of entering into a contract for the work, having spent only the minimum amount of money in the preparation of a single drawing. Clients often invite tenders using this system with no prior knowledge of building work and possessing little idea of the likely cost and inadequate financial resources for the scale of work proposed. However, for many small builders, this type of work will be the mainstay of their business and experience will hopefully give a contractor the ability to recognise the signs when the tendering circumstances and the client's intentions are not genuine.

2.11 Management contracting

Management contracting is a method of managing a project where the main contractor assumes full responsibility for the organisation and carrying out of the work on site. In return, he is reimbursed on a percentage basis or lump sum fee for his expertise in this field and for his overheads and profit.

The main contractor in charge of the project, whilst working in conjunction with the other members of the design team (architect, Q.S., consultants etc.), does not undertake any work directly – except for the provision of such items as temporary buildings and services, attendance on subcontractors, certain plant, insurances etc.

A central feature of this system is that all the work to be carried out is specified in separate work packages, each of which is then tendered for on a competitive basis by specialist sub-contractors. The documents comprising the work packages may be bills of approximate quantities, drawings, schedules or any other information considered appropriate.

Subcontractors once appointed, work under the direction of the main contractor as normal. Management-contracting is used to good effect on larger, complex projects where the special skills and expertise of the main contractor can be utilised in the areas of design management and evaluation, cost planning, programming, project administration etc.

The specific tasks undertaken by the main (managing) contractor fall, therefore, into two principal categories namely, (i) Sub-contractor control and (ii) design team integration, with regular site meetings taking place in order to resolve potential problems and ensure the smooth running of the work.

The advantages of Management Contracting may be summarised as follows:

(a) The main contractor has an early involvement in the design work and is able to advise on construction method etc.
(b) Client involvement is demanded all the way through the development process, thereby ensuring his satisfaction by

accommodating his requirements with regard to design changes as the work progresses. Information regarding the current financial position of the contract can be related to the client quickly where a close integration of the building team exists.

(c) Savings in time can be expected where a close co-operation exists between all parties.

(d) A high level of competition for the work may be achieved providing the client with good value for money. The main contractor can be selected by tendering on the basis of the fee and/or the provision of on site services and facilities. Sub-contract work (amounting to approximately 80% of the project) is tendered for competitively as suggested above.

(e) The system allows for flexibility. Design changes are easily accommodated since the 'work packages' need not be prepared until actually needed.

However, as may be expected, management contracting is not without a number of drawbacks namely:

(a) No firm price can be obtained by the client before the work begins as a great deal of investigative and design work remains incomplete.

(b) An extra tier of management is introduced. In addition to the main contractor's management staff on site, each sub-contractor would require his own organising and supervisory staff. Apart from the extra cost, this could lead to difficulties in communication and administration of the contract.

(c) A great deal of faith is invested in the main contractor's success in the monitoring, evaluation and control of the work executed in the sub-contract work packages.

The main contractor therefore needs to be selected with great care and must be fully competent and conversant with his legal and financial responsibilities.

(d) A large degree of success depends upon a much closer involvement by the quantity surveyor. With work being let in phases and the design evolving as the contract progresses, an up to date and accurate prediction of costs is continually required.

Summary

Method of selection	*Uses and main features*
Orthodox bill of quantities	Suitable for most projects provided all planning and design work has been completed at tender stage. Does not provide for contractor participation at design stage.

	Provides the client with good degree of control over financial aspects of the contract.
Bills of approximate quantities	Used mainly for civil engineering works and projects where detailed design work is incomplete at tender stage. Enables the client to make an early start on site. Rates in the bill are used to value the remeasured work.
Cost reimbursement contracts	Used where the extent and scope of the work cannot be ascertained at tender stage. Contractor is reimbursed with his actual costs plus a fee for overheads and profit. Encourages greater co-operation between client and contractor The pricing of variations does not pose a problem since the cost of *all* work done is recorded. Careful recording of costs is required; administration work is difficult.
Target cost contract	Used in same circumstances as above, but introduces incentives for the contractor to minimise costs.
Continuity contracts	Used where the client can take advantage of a contractor already on site in order to save time and money on a second project. Contracts usually entered into on an *ad hoc* basis. Original bill of quantities can be used as a basis for valuing new work.
Serial contracts	Similar to above, but the approximate number and size of similar future contracts are known beforehand. The successful contractor for the first project undertakes to carry out the remaining contracts using rates in the original bill of quantities to value the work.

Term contracts	Used for large-scale maintenance and repair work. Contract sum is unknown at the time of tender. Competition is based on a schedule of rates. Successful contractor undertakes to carry out all work given to him during a stated period of time, the work being measured and valued as it is completed.
Two-stage tenders	Used where the client wishes to involve the contractor in the planning and programming of the work. Suitable contractor initially selected on a competitive basis using a schedule of rates which are then used to value the work when final designs are complete.
Design and build	Used where the client wishes to take full advantage of a contractor's skill and expertise in a specialised field. Contractor carries out all design work, prepares his specification and price in addition to the building work itself. Usually gives the client an efficient and speedy service since the contractor is working to his own design.
Drawings and specification	Used for small works where the cost of preparing a bill of quantities is not justified. Presents the contractor with the additional task of preparing his own quantities. Terms and conditions of contract should be clearly stated to avoid confusion.
Management contracting	Used where the main contractor can use his managerial expertise in the design, planning and administration of the contract. The work is carried out entirely by specialist subcontractors operating under the direction of the main contractor.

Chapter 3

The preparation of a tender

Having determined the method by which the contractor will be selected and the necessary documentation having been prepared, the next stage in the process is the invitation of a number of suitable contractors to submit tenders. This chapter deals with the tender period from the contractor's viewpoint and is taken as being the time from the date of receipt of the tender documents by the contractor up to the date for submission of tenders some three or four weeks later. It deals mainly with the tasks the estimator has to perform during that time and how the final tender figure is produced and submitted.

3.1 The tender documents

In a further chapter the various sources of a typical contractor's business are considered, how a small to medium-sized firm looks to many different clients to provide him with work and how the size and complexity of the job determines, to a large extent, the nature of the tender documents. By 'tender documents' we mean all the relevant information about the proposed contract, rules, conditions, etc. supplied to the contractor which will enable him to price the work as accurately as possible, taking into account all the special peculiarities which every different building project possesses. The more detailed information which can be provided to the contractor at this stage, the better, since any ambiguity, lack of facts or confusion can easily lead to problems and disputes later on during the contract period.

The tender documents may include:

1 A covering letter pointing out specific important contractual details such as the contract period, starting date, firm or fluctuating price required, etc., together with details of any special problems to be encountered such as restricted hours of working, the likely delay in acceptance of a tender by the client and so on.
2 Bill(s) of quantities.
3 As many drawings as are available at this stage.
4 The specification.
5 Form of tender.

A sound estimate depends not only on the time allowed for completion and the quantities involved, but also on the method of carrying out the various building operations. An estimator must be able to judge for himself with the aid of specialist advisers (plant manager etc.) the most economical way of performing each task. He must be able to identify areas of work which can benefit from the extensive use of mechanical plant, for example, or where enforced labour-intensive operations must be carried out, where economies of scale can be achieved, etc.

The assumptions made by the estimator in this connection at tender stage will depend entirely on the accuracy of the information given to him by the architect and the quantity surveyor in the tender documents.

3.1.1 The 'estimate' and the 'tender'

Often, both these terms are used loosely to describe the same thing – namely the figure which is submitted to the client on the form of tender, i.e. the price for which the contractor is willing to carry out the work. The estimate and the tender, however, are different sums and can be defined as follows:

The 'estimate' is the net cost to the contractor of carrying out the work shown on the drawings and described in the bill of quantities and specification. It comprises labour, material and plant costs together with the cost of providing all the site services such as temporary telephone, site huts, transport, electricity, site supervision etc. The 'estimate' is exclusive of management costs, overheads, profit, allowances for risk and other tender adjustment required, and it is only when these additional sums have been considered and added on to the 'net cost' as defined that the 'estimate' is converted into the 'tender'.

3.2 General procedure

3.2.1 The arrival of the tender documents

Unfortunately, in the building industry a steady flow of work is rarely achieved and it must be accepted that wide variations in the level of enquiries will be common, contractors sometimes being engulfed with jobs to price not knowing which way to turn, and sometimes becalmed for weeks without a single enquiry appearing on the horizon!

If a regular workload could be sustained enabling a contractor to plan ahead and work more effectively, then perhaps he would be able to budget for a fair profit margin and, what's more important, be reasonably certain of achieving it!

The erratic nature of the flow of work in the industry invariably leads to high tender levels in times of plenty and massive price cutting in times of scarcity, with contractors becoming involved in fierce competition to gain what little work is available, often leading to the all too common situation of widespread bankruptcies.

Given the nature of the industry the estimator must somehow cope in all weathers. If clients have followed the recommendations of the 'code of procedure for single-stage selective tendering' in respect of circulating a preliminary enquiry for invitation to tender, the estimator will be aware of the impending arrival of the tender documents and so he can plan for this by setting aside some time in his diary. However, many enquiries arrive on the estimator's desk unannounced and in this situation, where a number of jobs arrive for pricing all at the same time whether expected or not, the estimator and management must decide which contracts to price and which must be reluctantly returned through lack of sufficient time to produce a tender.

Unfortunately, there is no standard time period allowed for a contractor to submit a tender; the code referred to above recommends a minimum of four working weeks but in practice the time allowed can be as little as a matter of days, depending on the size of the job.

In practice, during a busy period the estimator will be pricing perhaps three or four jobs at once, each one probably having a different tender period and a different date for submission of tenders. Therefore, having to work under this sort of pressure, the estimator must be selective in which jobs he prices, probably turning away those which on the face of it appear unlikely to go ahead. Unfortunately, not all enquiries received by the contractor are genuine and this can involve him in an expensive waste of time pricing abortive jobs. Some enquiries can consist not of a detailed bill of quantities which is very convenient to price, but of drawings and specifications which present their own special difficulties, not least of which is the length of time needed to price the work in addition to having to take off the quantities.

Since time is always limited for the estimator, he must use the short tender period to the best effect and plan his work very carefully if he is to produce an accurate a tender as possible by the date stated. (Again, in practice, even the most experienced estimator is often reduced to blind panic half an hour before the tender is due to be submitted, frantically ringing round suppliers and sub-contractors for last minute prices – not to be recommended!)

On receipt of the tender documents the tasks to be performed by the estimator can conveniently be divided into five main areas as follows: The decision to tender (3.2.2); The collection of information (3.2.3); The preparation of the estimate (3.2.4); The adjudication (i.e. the conversion of the estimate into a tender) (3.2.5); The submission of the tender (3.2.6).

3.2.2 The decision to tender

The contractor's estimating capabilities are a scarce resource, added to which the cost of producing a tender can run into hundreds of pounds and, bearing this in mind, he will probably take great care before committing himself to the costly estimating operation.

How does the contractor select those jobs he intends to price and those which he will return to the client with a polite note stating that 'on this occasion we regret that we are unable to submit a price'?

His first consideration no doubt will be the genuineness of the enquiry as stated earlier; beyond this, he would have to take into account such factors as those now listed.

1 The nature, size and complexity of the work

A large project promising perhaps two or three years' continuous work may seem very tempting. However, the realistic contractor, realising his limited resources of men, plant and equipment available may not wish to overstretch himself in tackling a large single contract. Indeed, even if he were successful in obtaining the contract he may well have to commit all of his resources to this one job thereby virtually excluding himself from the market for the next two years, his regular clients in the meantime having forgotten about him and found themselves other builders! Perhaps the type of work will not suit his workforce. If the contractor is a small, old-established firm with a reputation for high-quality joinery work, then perhaps a contract for a factory unit comprising concrete slab, steel frame and cladding would not suit his men and he would find it difficult to make a profit.

2 The principal parties involved

Who the client is and who the architect and quantity surveyor are may well determine the contractor's attitude towards the work.

Certainly, if he has had unfavourable dealings with any of the those parties in the past, whether in the realms of contract administration, final account agreement etc., it will colour his opinion of them and if serious enough could lead him to reject the enquiry. Alternatively, if there are good relations between a contractor and a particular client, then perhaps the contractor will be anxious to obtain the contract in the knowledge that disputes will be kept to a minimum resulting possibly in further work becoming available at the end of the job.

3 The value of the main contractor's own work, PC sums etc.

The contractor, being prepared to accept the responsibility of running the contract, will no doubt be looking for a reasonable amount of work which can be carried out by his own workforce. Undoubtedly he will have to sub-let some of the specialist trades to his own sub-contractors and the 'PC sum work' will be carried out by nominated sub-contractors. However, his first consideration is in keeping his own men employed and if the work fails to provide this basic need by involving mainly specialist trades, he will obtain little benefit in return for the headaches and administrative problems in trying to keep fifteen or twenty sub-contractors to their programme target dates!

4 The time allowed for tendering

The tender must be capable of being prepared with a reasonable degree of accuracy within the time allowed. This is a fundamental requirement which can only be determined by the estimator. Often a client will ask for a tender to be submitted in an unrealistically short period of time, probably because he wants to make an early start on site. This can result in the estimate being prepared in a hurry, often with approximate lump sums inserted against items bracketed together with the obvious greater possibility of errors being made.

5 Firm or fluctuating price required

It may be that a particular contractor is not prepared to accept the risk of pricing firm price contracts and will only be interested in submitting a tender on a fluctuating price basis. Not many contractors would be in the fortunate position of being able to select which jobs to price in this way, but in the event of the contractor finding himself having to make a choice between pricing a firm or fluctuating contract when busy, then perhaps he may choose the latter if only because it relieves him of a large degree of risk in attempting to predict what inflation is likely to be.

6 Contract details

Important factors such as the contract period, starting date, whether any phased work exists, access to site, working conditions etc., may all have some bearing on the suitability of the job and therefore should be discussed at this early stage by the management.

7 Current workload

It may happen that since the contractor first intimated that he would be willing to submit a price when a preliminary invitation to tender was received some weeks earlier, the situation could have changed. Perhaps, at the time the contractor was desperately short of work but by the time the tender documents had arrived he could have obtained maybe two or three large contracts and now finds himself with a full order book and unable to take on any more work for the time being. This being the case, he may very reluctantly be forced to return the tender documents.

All the above information would be recorded, probably on a standard printed form, by the estimator in order that the possibility of submitting a tender can be discussed. If, after due deliberation, it is decided not to submit a price, the tender documents should be returned promptly to the architect, but if a tender is to be submitted, then the next task is:

3.2.3 The collection of information

As previously stated, the time allowed for tendering is always limited and so the estimator must plan his programme very carefully, having only three weeks or so to complete his work. Before he can begin the actual mechanical process of pricing each individual bill item, a great deal of information needs to be gathered to enable him to price the work accurately. Material prices must be obtained, together with sub-contractors' quotations, plant and equipment hire rates, electricity and telephone charges, insurance premiums, water rates, etc.

It is important to give prospective suppliers and sub-contractors as much time as possible in which to prepare their own quotations; the more time which can be given to them to do this, then the more accurate their prices to the estimator will be; indeed, the accuracy of a sub-contractor's price incorporated into the main tender may make all the difference between the contractor himself being successful or failing to obtain the contract.

1 Enquiries to suppliers

The estimator will almost certainly use standard printed enquiry forms which can be adapted for use on any contract and sent out to suppliers and manufacturers. Although some prices will be obtained by telephone, the supplier will be requested to confirm his quotation in writing in due course, especially if copies of quotations have to be submitted in accordance with the conditions of recovery of fluctuations using the 'conventional' method (page 156). The source of supply of some materials may be restricted to a specified supplier if one or more are actually named in the bill of quantities, but for the majority of the materials, the estimator will choose a number of regular local suppliers from whom he can obtain prices for general items such as wall ties, damp-proof courses, nails, screws, polythene sheeting, timber, cement, aggregates, etc.

The information given by the estimator on the enquiry form to suppliers should be sufficient to enable an accurate price to be obtained. Such information would include those detailed under (a) – (f) below.

(a) The specification of the material. The type and quality of the materials involved should be clearly stated, the details being obtained from the bill description and specification. If this is lengthy, photocopies of the relevant pages of the bill and specification can be sent with the initial enquiry. Since one of the contractor's main obligations is to use only the standard of materials specified, it is obviously important for the estimator to make sure that on receipt of the quotations from suppliers, the prices are indeed related to the material to be used and not merely for a close substitute! What appears to be a very low price quoted by a supplier should be viewed with suspicion and a careful check made to ensure that there is no confusion over what is required.

(b) The quantity of materials required. The quantity of any one particular item required can often have a considerable effect on the price quoted by a supplier. A manufacturer will be anxious to take advantage of the opportunity of obtaining a large order by quoting competitive prices or offering discounts for bulk purchases since not only will a large turnover of stock be achieved with the minimum selling cost per unit, but also the handling and distribution costs per unit are significantly reduced, especially where an order from the contractor can be delivered to site in full lorry loads direct from the factory without having to be subsequently double-handled by a builders' merchant. If the estimator himself is keen to keep his tender as low as possible, then he may wish to take advantage of these trade discounts and incorporate these lower prices into his tender.

(It is of course dangerous practice to use the quantities in the bill to order materials without further reference to drawings etc., but for the purpose of obtaining quotations, using figures direct from the bill will suffice.)

Conversely, where only small quantities of a material are required, which may constitute only part loads, the price per unit can be significantly higher than the 'list price' due obviously to the opposite of the reasons stated earlier. Often, prices of materials are quoted by suppliers with the stipulation that the price stated will only apply where certain minimum quantities are purchased.

For example: A drainage materials supplier may quote standard list prices subject to an order having a minimum value of, say, £600.

For orders between £400–600, he may quote list price plus 5 per cent; for orders between £200–400, list price plus 10 per cent and so on until for small orders worth less than say £100, the customer may have to pay as much as list price plus 30 per cent.

The price of carcassing timber per cubic metre is often quoted provided the timber is purchased in minimum 1.25 cubic metre lots.

The quoted price of reinforcement per tonne may be applicable only when a minimum of 6 tonnes is ordered with deliveries less than this amount attracting a tonnage surcharge by the stockholder.

The price of cement is often quoted as being 'delivered to site in 10 tonne loads' and so on.

In order to recover the extra cost of purchasing small quantities it is important for the estimator to incorporate the higher prices within his tender and so he must be able to identify where his material costs will be higher than normal, since on occasions the wording drawing his attention to the fact that an extra for small quantities or part loads will apply may well be disguised in the small print on the back of the quotation!

(c) An indication of delivery date. In order to carry out the contract as efficiently as possible, careful co-ordination of men and materials is vital and so the contractor must plan ahead and organise deliveries of materials on to the site as soon as he starts work. Not only would the storage of so many items create problems, but also he would have difficulty in financing the purchase of all the materials at once, especially since the architect can refuse to include the value of goods in an interim valuation if they are brought on to the site prematurely. At the same time, however, the contractor must ensure that materials can be delivered in time to fit in with his planned programme of work, especially if clause 25.4.10.2 is deleted, barring the contractor from claiming an extension of time if he is unable to secure the necessary materials with which to complete the works by the stated date for completion.

Therefore, the estimator has to make sure that a prospective supplier is able not only to provide him with a competitive price, but also can effect delivery of the goods when they are needed on site. It is pointless using a low price in the estimate if the supplier cannot meet the required delivery date, resulting in the contractor losing money at a later stage by being forced to buy from an alternative supplier who can deliver the goods on time, but only at a higher price.

(d) The terms upon which the price is required. Ideally, the estimator would prefer the supplier to submit his quotation on the same basis as the terms of the main contract, particularly if it is a firm price basis! However, being realistic, the estimator must expect many suppliers to be unwilling to commit themselves to a fixed price

for any length of time; the maximum period for which quotations are open for acceptance, often being no more than 30 days from the date of the quotation. Thus, although the estimator may request a fixed price quotation (with tongue in cheek perhaps!), invariably the price submitted is qualified by the words 'prices ruling at date of despatch'. Often the estimator has no choice but to use a 'fluctuating price' for some materials in his 'firm price' tender which involves him in the risk of adding on a percentage to cover for likely increases. Should the terms of the main contract incorporate a fluctuation clause, then of course the problem does not arise, the contractor being able to recover increases in cost due to rises in the market price of materials as work progresses.

(e) Discounts and pro-forma. Although the estimator may not request it, discounts are often offered by some suppliers in return for prompt payment within so many days of the delivery date, the time allowed for payment and amount of discount being part of the terms of the quotation, 2½ per cent discount for settlement of the account within 30 days being common practice.

The estimator may, if he wishes, take this into consideration if he wants to submit a very keen tender, by reducing the price of such materials by the amount of the discount, although it is unlikely to be a good idea as it must be borne in mind that in order to obtain the discount, the invoice must be paid promptly, which may adversely affect the contractor's cash flow. It is more likely that the contractor will benefit more by taking advantage of any extended credit facilities available thereby avoiding paying the bill until the last possible moment!

The problems of financing the purchase of materials on site is compounded when a supplier insists on payment by 'pro-forma' method. This means that the contractor is invoiced for the goods as soon as he places his order, which will not be processed until payment is received. The contractor is unable to recover his outlay for the materials until they are delivered to the site or his workshops and included in an interim valuation, thus further affecting his cash flow. The estimator could take this into account when pricing and include a sum in the tender to offset these likely losses.

(f) Address and details of access to the site, etc. The address of the site must be given not only to enable the supplier to know where to deliver the goods, but also to enable him to calculate his delivery costs and to include them within his price. In some cases, suppliers operate a system of charging for delivery on a 'radial zone' basis, the country being divided up into areas or 'zones' relating to their distances from the place of manufacture. The further the site is away from the point of manufacture, the higher the delivery charge.

Details of the nature of the site, access and unloading

requirements should be given to the supplier as these could be important if there are restrictions regarding height of vehicles, narrow site entrance, waiting regulations etc. In a busy town centre with the site surrounded by other buildings, the supplier may not be able to deliver in large articulated vehicles and may have to restrict deliveries to 5 tonne tipper lorries, thereby incurring additional cost.

Regarding methods of unloading, a supplier may be able to offer other alternatives in addition to the normal way of off-loading by hand. For example, with block deliveries, a vehicle may be equipped for off-loading by a crane attached to the lorry or the blocks could be palleted for unloading by fork-lift truck. For restricted sites such as high-rise developments in city centres, a clamp service may be available, which is a device capable of being attached to the hook of the site crane enabling large packs of blocks to be lifted directly from the lorry to the point of use.

The preferred method of unloading should be stated clearly on the enquiry form since each one could produce different costs.

(g) The date by which the quotation is required and period for which it is open for acceptance. Obviously the date by which the quotation is required by the estimator should be stated in order to give him sufficient time to be able to check it thoroughly and raise any queries if necessary before using it in his tender.

As stated earlier, however, although the estimator would indicate the period for which the quotation is required to be open for acceptance, individual suppliers often adhere to their own policy regarding fixed price quotations (see paragraph (d) above).

2 Enquiries to sub-contractors

In addition to collecting prices relating to materials, the estimator will also need quotations for work which will subsequently be sub-let. Most contractors rely on smaller sub-contracting firms to carry out work of a specialist nature on their contracts and, depending on which trades are carried by the main contractor, work frequently sub-let can include felt roofing, plumbing, plastering, structural steelwork, specialist flooring and ceiling finishes and tarmacadam work. It must be remembered that before any work can be sub-let, the contractor must obtain the consent of the architect to each prospective sub-contractor and so the estimator must be satisfied that the reputation and standard of workmanship of each one is at a reasonable standard and will be acceptable to the architect.

The estimator will have a number of regular sub-contractors for each specialist trade to be sub-let, firms the contractor usually deals with and where good relations exist between them. Perhaps four or five sub-contractors for each trade will be chosen, involving as many as forty separate enquiries, and again, as in the case of prospective suppliers of materials, the enquiries to sub-contractors should incorporate a number of important points detailed in (a)–(e) below.

(a) The items to be priced. The bill items to be priced should be clearly stated. In most cases copies of the relevant pages of the bill of quantities and specification can be used for this purpose and should accompany copies of any drawings if applicable. A great deal of care must be taken to ensure that each sub-contractor receives exactly the same information since the omission of say the page containing all the sanitary appliances from the plumbing work section could result in one sub-contractor's quotation being considerably less than his competitors, bringing problems later on if this price is used in the tender without being checked.

Mistakes of this nature can be costly if undetected and can easily be made especially in view of the fact that the task of preparing the information to be sent out to be priced usually falls on a junior member of staff in the estimating department who is assigned to the long, laborious job of photocopying seemingly unending pages of the bills of quantities and specification.

(b) The terms and conditions of the main contract. Regardless of the private arrangement which may be made between a contractor and his domestic sub-contractors, payment for work which is sub-let will be made in accordance with the terms and conditions of the main form of contract. The estimator, therefore, will be anxious for each sub-contractor to commit himself to the same sort of risks as the main contractor himself will have to accept, which on occasions may not be possible as some sub-contractors will be unwilling to quote firm prices in times of uncertainty. The estimator must account for such deviations from the main set of conditions by adjusting the quotations of these sub-contractors where applicable to allow for extra costs and obligations not provided for.

(c) Details of anticipated starting and completion dates. In addition to the commencement and completion dates for the main contract being stated, the anticipated starting and completion dates for each sub-contractor should be given. These will vary according to the stage in the construction process at which each particular trade will be required on site and so provisional dates, taken from the preliminary programme (see later), will enable the sub-contractor to plan ahead and see if he will be able to cope with the work at the appropriate time, assuming an anticipated workload and his quotation being successful.

Sub-contractors carrying out later trades such as floor tiling, suspended ceilings, tarmacadam work, etc. are at greater risk. It may be that they will not be required on site for a further nine or ten months from the date the main contractor starts work, thereby having to calculate the effects of a year's inflation on their current costs. This being the reason why so many sub-contractors are reluctant to quote firm prices for any considerable length of time, and understandably so.

However, by giving them details of a likely starting date, the estimator at least is giving the sub-contractor an opportunity of making due allowance for inflation, even if subsequently he fails to obtain the firm price he may be seeking!

(d) The plant, materials, services etc., to be provided by the main contractor. A sub-contractor's price, which on the face of it looks competitive, may not be so attractive once the qualifications often contained in the small print on the back of the quotation are taken into account. Frequently, sub-contractors require the main contractor to provide, free of charge, a variety of services, plant, equipment, etc., the cost of which when taken into account by the estimator often relegates a particular quotation to the back of the queue! Examples of such services which can sometimes be required by a sub-contractor include:

- The unloading and storing of materials and plant by the main contractor, the sub-contractor often not being present on site when the delivery of his equipment is being made.
- The provision of a mixer, dumper and hoist free of charge for the mixing and laying of, say, floor screeds, for example, on upper floors.
- The free use of electricity and provision of cables and extension leads for the use of power tools.
- The provision of sand and cement for bedding floor tiles or in roofing works.
- The provision of a special mobile tower scaffold for fixing suspended ceilings or carrying out glazing at high level.
- The provision of a hardstanding comprising a bed of consolidated hardcore for a sub-contractor's crane to operate from when unloading and erecting structural steelwork or a pre-cast concrete floor. . . . etc.

Therefore, the estimator must identify those quotations which include for such items and those which need adjusting to allow for the cost to the contractor of providing these additional facilities. The likely cost being calculated by the estimator and added on to the respective quotations.

(e) The date by which the quotation is required and period for which it is open for acceptance. Such details would be given for the same reasons outlined in 'enquiries to suppliers' as would be the case also for providing details in respect of: *(f) Address and details of the site.*

3 Visit to the site and architect's office

Having despatched all enquiries to suppliers and sub-contractors as quickly as possible, the estimator will have a period of a few days before prices begin to come back. He can make good use of this time to collect further information, adding to his overall picture of the job by visiting the site and possibly the architect's office if necessary. The site visit is essential because however detailed the drawings and bills of quantities may be, they can often mean very little to the estimator until he has actually seen the place where the building is to be erected. It is his task to translate the information given on the drawings and in the bill in terms of cost, and what could be assumed to be an innocent, straightforward operation when viewed from the office desk may become a frustrating, complex and costly piece of work when considered in terms of the reality of site conditions.

He may visit the site by himself or, preferably, with the contracts manager who can advise on the practical working conditions likely to be encountered. Between them they can discuss and exchange opinions on construction methods, site layouts and other anticipated problems, thus establishing the basis for a report which may well influence the contractor's willingness to proceed any further with the venture! The information the estimator is seeking from his site visit would include the following items detailed in (a)–(f) below.

(a) Details of site ground conditions, spot items, etc. If the bill of quantities has been prepared in accordance with the sixth edition of the *Standard Method of Measurement*, detailed information in connection with the soil description, including water levels, bore holes etc., should be given. This makes the establishment of the level of efficiency of mechanical plant involved in the excavation work less risky than it used to be, especially since the water level can subsequently be re-established at the time the various excavation works are being carried out, thereby providing for additional costs to be recovered where subsequent variations in the water table affect the excavations. Even so, the estimator must still be able to judge the output of his mechanical plant by examining the conditions of the ground since the bill will not indicate where machines will be unable to work and where hand digging must prevail (for example, a drain trench required in close proximity to an existing unstable boundary wall or in the vicinity of underground services). Spot items, such as the demolition of an old air raid shelter or greenhouses, the grubbing up of hedges and cutting down of trees, filling in old disused manholes etc., can only accurately be priced by looking at the item on site. It is impossible to tell by reading the description in the bill whether the demolition of an old wall can be achieved by merely leaning on it or whether it requires a JCB or a compressor to do the job!

(b) Nearest available tip. In order for the contractor to work efficiently the site must be kept as tidy as possible with heaps of accumulated rubbish and debris being reduced to a minimum by being cleared away at regular intervals throughout the progress of the works.

Furthermore, any excavated material which is not subsequently backfilled around the foundations or used elsewhere on the site for making up levels or in landscaping work must be removed from the site. It is the contractor's responsibility to locate a tip and arrange disposal of any surplus spoil and waste materials using either a private or local authority tip where perhaps a land fill project is in operation. Often a charge per load is levied by the owner of a tip for the disposal of rubbish and this if applicable, together with the distance of the most suitable tip from the site, must be established in order for the estimator to calculate the cost of removing surplus material from site (see later chapters for detailed calculations).

(c) Security. On taking possession of the site, the contractor becomes responsible for all security aspects, keeping the building safe and secure each night, protecting his stockpile of materials, plant etc., with breaches in security often proving very expensive to him or his insurance company. Building sites are seen as easy pickings and attract all sorts of vandalism and theft with items such as copper tubing, packs of facing bricks and ironmongery proving popular targets for the petty thief. The location of the site itself would determine the extent of temporary protection required: obviously, if the site is situated away from built-up areas in a quiet country location the problem of vandalism may not arise and a small compound to house a few items of plant and equipment may be adequate, but at the other end of the scale, sites located in a run-down inner city area may be particularly vulnerable and need surrounding completely by hoardings or chain-link fencing topped with barbed wire and having sturdy lockable gates and guarded by nightwatchmen or security patrols. The cost of providing such temporary facilities would be priced by the estimator in the preliminaries section of the bill.

(d) Location of services – electricity, water, gas, telephone, etc. The provision of these services is essential not only for the operation of the completed building, but also on a temporary basis for the contractor's purposes during the construction, electricity being needed for site lighting during winter months, power tools, etc.; water for building purposes and washing facilities and telephone for contact with head office, suppliers, sub-contractors, etc., and so on.

It is the contractor's responsibility to provide such temporary services and the estimator must allow in his tender for the cost of installation, maintenance and removal on completion. The distance of the mains supplies in the road or footpath from the anticipated supply points on site will determine a large part of the expense involved. It is important therefore for the estimator to locate the position of existing services not only for this reason but also because there may be electricity cables and gas mains running under the site itself which must be determined before any excavation begins for safety reasons. Often, the exact position of old services are not shown on any drawings, with even the electricity and gas boards themselves being reluctant to confirm their existence! In the event of uncertainty, the estimator may decide it prudent to allow for the extra expense of hand digging in the vicinity of any underground services if no mention has been made in the bill descriptions. The local authority and statutory bodies would have to be contacted in order to establish the costs of tapping into the mains, connecting drains from temporary toilets into the sewer, the charge for constructing crossovers etc., and addresses and telephone numbers of the local depots and branches should be noted for future use.

(e) Access to the site and possible layouts. These factors have been mentioned earlier in relation to sub-contractors' and suppliers' enquiries and the estimator must now view the situation with regard to his own operations. It would be folly, for example, to base the pricing of the concrete work on the use of ready-mixed concrete if the entrance to the site subsequently proves to be too narrow or too low to allow the vehicle on to the site!

Where the entrance to the site must cross over an existing footpath, adequate protection of the public must be provided using hoardings, pedestrian walkways, barriers and lighting, etc. In addition, it is likely that the road kerbs and footpath itself will suffer from the constant passage of site vehicles during the course of the contract even though ramps of timber railway sleepers are often used to minimise the damage. If the tender is successful, the estimator may wish to take photographs of the vicinity of the site entrance, and agree with the architect beforehand the extent of any existing damaged areas since the contractor will be required, at his own cost, to reinstate any damaged work to its original condition on completion of the work.

If the site is in a busy town centre and delivery lorries are unable to get on to the site, the cost of unloading and stacking materials will be greatly increased especially in situations where even a dumper or fork-lift truck are unable to be utilised and labour only must be employed to do the job. Where a site crane is planned, however, it may be possible for materials to be unloaded and hoisted into position direct from lorries waiting in the street. The merits of each alternative operation must be carefully weighed and costed by the estimator in discussions with the contracts manager before a decision on how to tackle the problems will be made.

Thought must also be given at this stage to the time when the work is well under way and the site is full of activity with men, plant and machinery busy and daily deliveries of materials being made. Possible site layouts must be considered and planned to allow the maximum amount of free movement about the site, the choice of location of the site office and storage huts, stockpiles of bricks, aggregates and timber, etc. depending on the space available in relation to the building, external works, spoil heaps, etc. The decision made will have an important bearing on the future efficient running of the job.

The estimator, having completed these preliminary tasks will be starting to build up an overall picture of the contract in his mind, the size and scope of the work, resources needed to carry out the construction and the individual problems and difficulties associated with every contract which at this stage are so easy to overlook, but which could prove to be very costly. His initial thoughts even at this early stage will probably already be influencing the way in which he intends to price the bill items. The tender period is already a few days old, the estimator still has not put pen to paper with regard to working out any final prices; however, the foundation for a sound tender has been laid and this must be developed further before the calculation of any unit rates can begin. Prices for materials and sub-contractors' quotations will probably start to trickle in by this time and while these are being carefully examined for sufficiency, arithmetical errors and so on, the estimator can press ahead with:

(f) Preliminary meetings; Determination of construction methods; Preliminary programme; Cash flow forecasts etc. The estimator need not be a recognised 'expert' on every aspect of construction in order to be successful. The secret of efficient and profitable construction is teamwork and the estimator often relies heavily on advice and suggestions from the contracts manager, plant manager, joiners' shop foreman and so on, each having a valuable contribution to make in lending his own special expertise and experience to the preparation of the tender.

The project will be discussed at meetings with the other managers in order to determine the plant and other services required. Sequences of work, site layouts and site staff needed will be other items on the agenda. In most cases, the contract period is predetermined by the client and this is often the critical factor in determining how the job will be tackled. Sometimes, the contract period is somewhat less than the contractor would have liked and so he must be flexible in his approach and be prepared to adopt alternative methods of construction in order to complete the work on time. Often this involves him in additional cost where expensive items of plant must be used to increase output or where non-productive overtime must be worked, all of which must be taken into account in the compilation of the tender.

The contracts manager must examine his labour resources and decide, in the light of current commitments, whether the labour requirement for the job could be met by men already employed on the firm's books or whether it would be necessary to recruit additional labour from the locality or from further afield which would involve the extra cost of travelling and fare allowance (a typical calculation of this cost is included in Chapter 6, 'Preliminaries').

The plant manager would be able to advise on the items of plant and scaffolding which would be available. Where existing plant is spoken for on other contracts, capital outlay for the purchase of further equipment may be necessary which may well be the subject of a policy decision made only by the directors. If, having considered the labour and plant, it is decided that existing resources are rather scarce, the estimator may decide to sub-let further work which initially he had planned would be done by his own firm. It may be advantageous to sub-let the scaffolding, say, to a firm of specialist scaffolders, or the whole of the excavation work to a groundworks sub-contractor thereby releasing valuable men and equipment for other contracts. The estimator must consider not only the job currently being priced but the whole of the firm's operations which may include a dozen or more other contracts all in various stages of completion.

In an ideal situation his job would be made much easier by an endless supply of the right type of plant, equipment, craftsmen and labourers, but being realistic he must price the contract with his company's financial and technical limitations firmly in mind and work within the framework imposed upon him.

Further progress now having been made, a preliminary programme can now be prepared. A simple bar chart type plan may suffice and for the first time the proposed progress of the contract can be followed in diagrammatic form. An example of a typical bar chart programme is shown in Fig. 3.3, and although it may be subject to many amendments later on, at this pre-contract stage the important information it should contain would include:

1 The contract starting and completion dates.
2 Work which is to be tackled in stages.
3 The length of time alloted to major areas of work and their relationship to each other.
4 Trades which are to be sub-let to domestic sub-contractors.
5 Work which will be carried out by nominated sub-contractors.

The estimator should now be fairly clear about the nature of the work, how it will be tackled if successful, and the plant, labour and other resources which will be required. Armed with this information

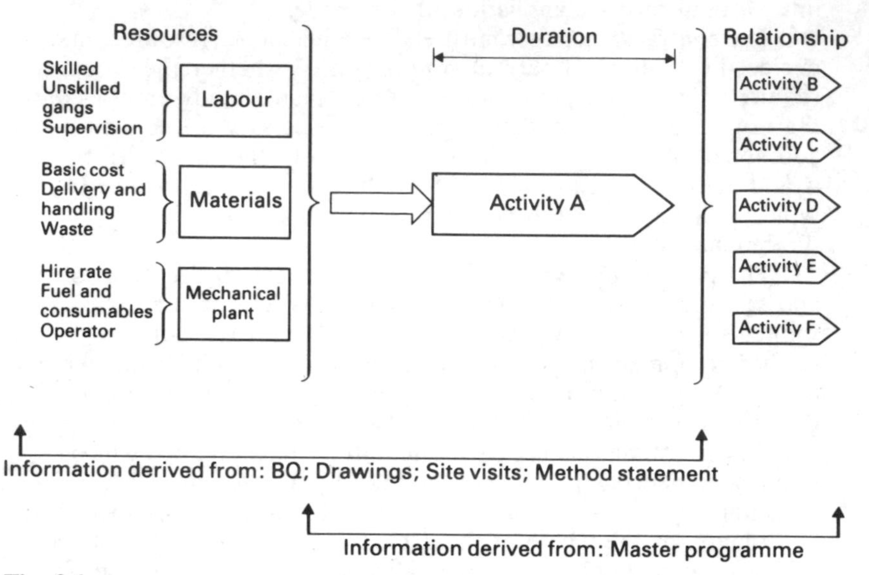

Fig. 3.1

and his suppliers and sub-contractors' quotations, he can now sit down and actually begin the preparation of the estimate – the next stage in the preparation of the tender. Despite the assistance given by others, however, the estimator must have in his mind a clear idea of the factors which influence the cost of building operations. The cost of any activity is dependent not only upon the nature and quantity of the work to be undertaken but also on the physical and contractual conditions under which the work is executed. Figure 3.1 illustrates how the cost of an activity is determined by:

1 The resources needed.
2 The duration of the operation.
3 The relationship between the operation and other activities.

In order to reduce the margin of error to a minimum when estimating costs, a great deal of reliable data is required from a variety of sources, namely the bill of quantities and drawings, information derived from site visits, and from the production of a **Method statement** in addition to the preliminary programme mentioned above.

The method statement

The method statement, an example of which is shown, is a 'statement of intent' by the contractor describing the methods by which he will tackle the work should his tender be accepted and showing the type and quantities of labour, plant and materials

Method statement

A. JONES and SONS (Builders)
CONTRACT: *St. Stephens' School*
SHEET NO: *2 of 6*

DATE: *1.10.84*
PREPARED BY: *R. Jones*

Substructure

ITEM	METHOD	QUANTITY	LAB.	PLANT	MAT.	TIME
A. Site strip and reduce level excav. and disposal	Wheeled excavator/loader – part tipping into lorries and transp. to council tip 5 miles. Part loading into dumper and depositing in temp. spoil heaps average 100 m distant.	700 m^3	2 No. excav. drivers; 3 No. lorry drivers; 2 No. dumper drivers 2 No. labs.	2 No. – type A excavator 3 No. 20 t tipper lorries 2 No. 2·30 m^3 dumpers	–	4 days
B. Found. trenches and drain trenches	Wheeled excavator with ½ m^3 bucket – load and remove to spoil heaps as for item A.	300 m^3	1 No. excav. driver; 2 No. lorry drivers; 1 No. dumper driver 1 No. banksmen	1 No. type B excavator 2 No. 20 t lorries 1 No. 2·30 m^3 dumper	–	3 days
C. Backfill	Dumpers and consol. by hand. Using mech. punners.	175 m^3	2 No. dumper drivers; 2 labs.	2 No. 2·30 m^3 dumpers 2 No. punners	–	2 days
D. Hardcore bed.	Using limestone D/d to site in 20 t lorries. Spread and levelled by JCBs and gang. Consolidate with 6 t roller. Bed thickness 225 mm.	1100 m^2	1 No. driver; 1 No. roller driver; 4 labs	1 No. type C loader/excav. 1 No. 6t roller	500t (allowing for compaction)	6 days
E. Concrete founds.	Using RMC poured direct into trenches – gang spreading and levelling in position.	250 m^3	4 labs.	–	RMC type X 262 m^3 allowing for waste	9 days
F. Breaking out old concrete slab	Compressor and gang – remove to tip using 10 t lorry.	60 m^2	1 lorry driver; 3 labs.	2 tool compressor 10 t lorry	–	5 days

Fig. 3.2

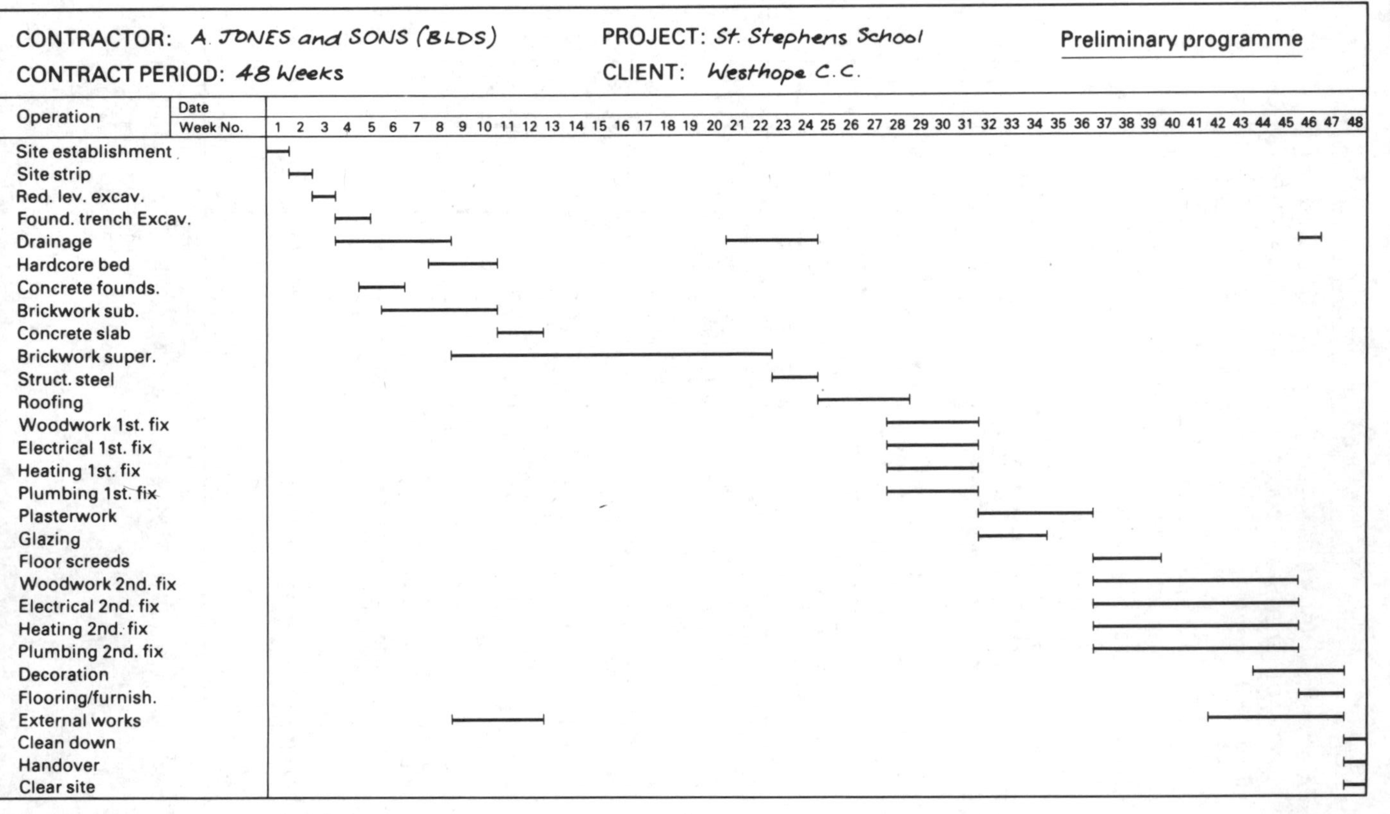

CONTRACTOR: A. JONES and SONS (BLDS)
PROJECT: St. Stephens School
Preliminary programme
CONTRACT PERIOD: 48 Weeks
CLIENT: Westhope C.C.
Operation
Date
Week No.
1 2 3 4 5 6 7 8 9 10 11 12 13 14 15 16 17 18 19 20 21 22 23 24 25 26 27 28 29 30 31 32 33 34 35 36 37 38 39 40 41 42 43 44 45 46 47 48
Site establishment
Site strip
Red. lev. excav.
Found. trench Excav.
Drainage
Hardcore bed
Concrete founds.
Brickwork sub.
Concrete slab
Brickwork super.
Struct. steel
Roofing
Woodwork 1st. fix
Electrical 1st. fix
Heating 1st. fix
Plumbing 1st. fix
Plasterwork
Glazing
Floor screeds
Woodwork 2nd. fix
Electrical 2nd. fix
Heating 2nd. fix
Plumbing 2nd. fix
Decoration
Flooring/furnish.
External works
Clean down
Handover
Clear site

Fig. 3.3

needed to carry out specific operations. The amount of detail recorded on the method statement whilst normally fairly brief, is of course at the discretion of individual contractors – indeed if the estimator wished, the whole project could be priced directly from the method statement if rates were to be inserted where appropriate. It's unlikely though that this would in fact be done although for significant sections of work, the pricing of the corresponding part of the method statement can provide a useful check against the pricing of the bill items.

In addition to establishing the principles upon which costs are determined, the method statement has a second use which may be seen as essential for those working on the project once the contract begins, by providing the guidelines within which the work must be undertaken in order to keep within the cost limits previously calculated.

3.2.4 The preparation of the estimate

If the 'spadework' involved in collecting the necessary information has been done thoroughly, the preparation of the estimate should be little more than a 'mechanical' process. The steps involved in this process are dealt with in detail in the following chapters, but for the moment can be summarised as follows:

1 The establishment of 'all-in' rates for key items such as labour, gang costs, plant rates, etc.
2 The use of these 'all-in' rates together with prices per unit for materials in order to calculate unit rates for each item in the bill.
3 Once the 'measured' items within the work sections have been priced, the value of the preliminaries can be determined.

In order to arrive at a figure for the 'estimate', i.e. the predicted net cost to the contractor of carrying out the work both in terms of the workmanship and materials used directly in the construction of the building and all the back-up of ancillary site services and supervision etc., two further sums must be added. These relate to:

1 The value of the PC and provisional sums contained in the bill of quantities together with any general and special attendance items which may have been priced; and
2 The value of domestic sub-contractors' quotations for work which is planned to be sub-let, suitably adjusted to allow for discounts (if any), attendances, conversion to firm prices, etc., as previously discussed.

To summarise, the estimate will comprise:

1 The net cost of the contractor's own measured work.
2 The net cost of the preliminaries section.
3 The value of PC and provisional sums together with attendances.
4 The value of domestic sub-contractors' quotations.

In order to convert the 'estimate' into the 'tender', the contract as a whole must be considered from a number of different viewpoints leading to adjudication.

3.2.5 The adjudication

In simple terms, the adjudication, or consideration of the tender, by the management involves adding to the estimate a further element for overheads and profit, and whilst on the face of it this may appear to be merely an academic exercise, in practice the final decision regarding the tender level should be made only after careful appreciation by management of all relevant factors affecting the company's well being. Whereas the estimator can work within his own narrow sphere of operations quite satisfactorily when pricing a contract, the management of the firm must take a much broader view covering the whole range of the company's activities which may involve other interests besides building.

Often a contractor will develop offshoots from the parent company setting up a subsidiary plant hire company or concrete block manufacturing department, etc., the relative success or failure of which may influence the final tender figure of the job now under consideration. If other companies within the group are performing well whilst the building company is struggling, the pressure upon the management to obtain work may not be as great as would otherwise have been if building had been the only business. In such circumstances the management may well be able to afford the luxury of avoiding cutting profit margins to the bone in order to keep a steady workload, safe in the knowledge that even if the building side of the operations was working well below capacity, the success of the other enterprises could keep the company afloat – at least for a time anyway!

However, most building contractors do not possess the security of such a fall-back position and must maintain a relatively full order book to stay in business. In such cases the final decision regarding the profit margin to be included cannot simply be a matter of adding the usual 10 per cent or so regardless of the consequences for although the policy and marketing objectives of the firm should set the initial tender level, the management must be sufficiently flexible in their approach to be able to recognise and adapt to changing circumstances. With this in mind it is not surprising to see profit margins varying from one job to another depending on the situation at the time and it would be wrong to assume that it was the result of erratic and inconsistent management. Often, days can be spent involving wide-ranging discussions before a final decision on the profit level is determined.

The problems in settling the tender figure by the management is not made any easier by a number of uncertainties common to all building contracts. They cannot, for example, be certain that the estimator has not underestimated his costs or overestimated the capabilities of his workforce and mechanical plant; for a large part of the contract period, progress is often determined by the vagaries of the weather; the contract may be hindered by plant breakdowns, difficulties with sub-contractors or even by bad site management. However much management may wish to be covered for all eventualities there is a limit to how much the tender can be 'loaded up' if the price is to remain competitive. Other factors which no doubt need to be considered as having some bearing on the final tender level may include those detailed in items 1–4 below.

1. The workload

The situation can change very quickly in building and whereas a contractor may have been very anxious to price a new job, having little or no guaranteed work ahead of him when the tender documents arrived, perhaps now some four weeks later when the tender is due for submission he could have picked up one or two other jobs within this time thereby relieving the pressure upon him. He may now not be too concerned whether he gets the job or not; indeed, it may pose further problems if he is successful since the acquisition of another contract would stretch his resources to the maximum, endangering his whole scheme of operations. In such a case it is likely that the management would add a higher than normal profit element, being delighted if they unexpectedly pick up the job or, as is more likely, not too concerned when the tender is lost.

At the other end of the scale, when perhaps other expected work fails to materialise and the prospect of making part of the workforce redundant looms large, then the contractor may reluctantly be pressured into pricing the work with only a nominal figure for profit or worse at no more than cost!

2 Market level of competition

During the course of pricing the work, the estimator will generally discover who his competitors are by one means or another thereby giving him an indication of the level of competition.

The keenness of the competition can be gauged by looking at the past records of the other firms pricing their work, their reputation and the extent of their current workload. Much of this can only be a matter of speculation of course, but knowing that only a few competitors exist of whom one or two are likely to drop out may tempt management into thinking it safe to increase the profit margin.

3. The project reports

By considering the project reports prepared by the estimator, the conditions of the contract, terms of the quotations from sub-contractors and suppliers, etc., the contractual and construction risks can be assessed. The extent to which the contractor expects to be rewarded for accepting such risks will be largely a question of company policy but, generally, the greater the degree of risk and uncertainty involved in the contract, the greater the profit margin that will be expected by the management. From the company's and shareholders' point of view, a guaranteed return can be obtained from 'safe' investments such as bank deposits and gilt-edged securities. Where an element of risk is attached to an investment, it follows that a higher rate of return would be required to make it worthwhile.

4. The resources required for the project

It may happen that a current contract will be shortly coming to an end which will release a number of technical staff, foreman, agent, quantity surveyor, etc., and other resources such as site huts, plant and equipment. If the timing is right, it would suit the contractor to move smoothly from a successfully completed job to a new one immediately which, if existing resources are used, could save a great deal in capital expenditure. However, on the other hand, where all plant, scaffolding, etc. is already employed on other sites, further capital expenditure in this direction may not be a good idea especially where the likelihood of picking up more work in the near future looks bleak. The danger of 'over trading' is often present with the possibility of increased turnover very tempting. However, once a contractor becomes overstretched both financially and in terms of technical and management staff, the scramble to recover ever-increasing overhead costs often results in price cutting to secure the necessary volume of work, leading inevitably to disaster.

The profit level thus becomes the most variable element of all, subject to being squeezed and stretched from one job to another and from one month to another depending on the economic and sometimes political climate at the time. Management therefore must always be aware of potential changes in the trading position of the company and be prepared to adjust profit margins accordingly.

Whereas the percentage to be included in the tender for profit is variable, the overhead percentage is more likely to be fixed.

The establishment of an overhead percentage is illustrated in a later chapter and shows how a typical contractor determines the amounts of his overheads and sets out to recover the sum involved throughout one financial year.

Job.	Alterations and Extensions – Southill School
Client	Sanderside
Date for Submission	10.10.84

TENDER SUMMARY

		net bill total	£ 452,604
LESS			
provisional + p.c. sums	£ 182,000		
private subcontractors	£ 101,585		
contingencies	£ 10,000		£ 293,585
		main contractor's work	£ 159,019
ADD			
preliminaries			£ 35,726
			194,745
overheads	@ 8%		£ 15,580
			210,325
profit	@ 5%		£ 10,516
			220,841
provisional + p.c. sums	£ 182,000		
private subcontractors	£ 101,585		
contingencies	£ 10,000		£ 293,585
			514,426
bond	@ .075%		£ 3,858
			518,284
water			£ 1,000
ADD/~~DDT~~ tender adjustments	(estimate of shortfall in fluctuations)		£ 4,000
		TENDER TOTAL	£ 523,284

SUMMARY OF PRIVATE SUBCONTRACTORS USED IN TENDER

name	trade	bill ref.	quotation	discount
T. Southwall & Sons	Demolition	2/1 A–Z; 2/2 A–X	10,500 net	–
Maron ltd.	Asphalt work	2/12 A–T	15,364 – 2½ %	384
S. Smith & Son Co.	Tile roofing	2/18 A–S	12,520 net	–
Cliveden htg. Co.	Plumbing	2/70 A–X; 2/71 A–Z; 2/72 A–H; 2/112 A–C	25,972 – 2½ %	649
M. Dennis	Plastering	2/80 A–W; 2/81 A–X	14,856 – 2½ %	371
Caroline carpets	Flooring	2/82 A, B; 2/84 C–J	6,702 net	–
Jimclaire ltd.	Tarmac	2/115 AR	15,671 net	–
		£	101,585	1,404

SUMMARY OF P.C. SUMS

operation	bill ref.	amount	discount
			(all – 2½ %)
Electrical work	3/1 A	55,000	1,375
Heating inst.	3/1 F	42,000	1,050
Structural steel	3/2 B	20,000	500
Fire alarms	3/2 G	25,000	625
Furniture & desks	3/2 M	30,000	750
Landscaping	3/2 T	10,000	250
	£	182,000	4,550

Fig. 3.4 Standard form of tender summary

3.2.6 The submission of the tender

Having determined the amount to be included in the tender for profit and overheads, the estimator can then compile his final 'tender summary'. Most contractors use variations of a standard printed form for this purpose, a typical example is shown in Fig. 3.4.

Notes to Fig. 3.4

(a) Brief details of the contract recorded, particularly the date for submission which should always be foremost in the estimator's mind. If the tender is submitted after the stated deadline, it is likely to be rejected and a great deal of work could have gone to waste.

(b) It is assumed in this instance that the contractor prices overheads and profit on the value of his own work only and is quite happy to include sub-contractors' quotations within his tender at net cost. The value of the main contractor's own work is found by deducting from the net bill total obtained, the amounts for PC and provisional sums, domestic sub-contractors and contingencies.

(c) Since the contractor will almost certainly provide all the services called for in the preliminaries, this will be classed as his 'own work' and therefore added to the value of own measured work in the bill.

(d) A predetermined percentage to cover overheads will be added to the net cost of the preliminaries and measured work.

(e) A profit margin is then added which may be either a straightforward percentage based on the value of the work involved or a lump sum depending on the policy adopted by management for this project.

(f) The PC sums, domestic sub-contractors' quotations and contingencies are added back in at net cost.

(g) Since the sums to be included for the bond (if required) and the provision of building water by the local water authority are based on the total value of the contract, their amounts can only be assessed at this late stage when the tender figure is known.

The final figure having been thus determined and confirmed by the management, all that remains is for the tender to be conveyed to the client usually on a standard form of tender, a typical example of which is reproduced in Fig. 3.5.

The tender figure is written in both words and figures and the form signed by the contractor. At this stage, the priced bill of quantities is not submitted and will only be requested by the client if the tender is the lowest when it will be submitted for checking to the client's quantity surveyor.

FORM OF TENDER
COUNTY OF NORTH SANDERSIDE

TENDER

for

ALTERATIONS AND EXTENSIONS

at

SOUTHILL SCHOOL

TO: The Council of the County of North Sanderside

~~I~~/We the undersigned offer to execute the above mentioned works in accordance with the Drawings, Bills of Quantities and the Conditions of Contract for the sum of:-

Five hundred and twenty three thousand two hundred

(a) and eighty four pounds NIL pence

£ 523,284 : 00 p

~~I~~/We further undertake:-

(b) That this tender will remain open for acceptance by the Employer for a period of three months from the date of the tender.

(c) That where the Tender sum exceeds £100,000, jointly with the Surety mentioned below, to enter into a Bond in the form prescribed by the County Council in the Bills of Quantities in a sum equal to 10% of the Tender Sum for the due and proper performance and completion of the Works.

The following Company has agreed to act as Surety and to enter into a Bond in the prescribed form in the event of this tender being accepted.

NAME OF COMPANY Worldwide Insurance Co. Ltd. London

~~I~~/We confirm that the full cost of providing such Bond is included in ~~my~~/our Tender.

(d) ~~I~~/We hereby warrant that, to the best of ~~my~~/our knowledge and belief, ~~I~~/We have complied with the General Conditions of the Fair Wages Clause of the Conditions of Contract for at least three months immediately preceding the date of the tender.

(e) Signed R.J. Taylor

Designation ~~Director~~/Company Secretary/~~Agent~~ authorised to enter into Contracts
(Delete as appropriate)

on behalf of:-

NAME AND ADDRESS OF CONTRACTOR A. Jones (Blds) Railway Works, Southill

Telephone No 44621 Date 9/10/82

Fig. 3.5 Form of tender

IMPORTANT

(f) 1. Should the Contractor wish to be invited to tender for work covered by Prime Cost Sums in the Bills of Quantities this should be indicated below:-

PRIME COST SUMS FOR WHICH THE CONTRACTOR DESIRES TO TENDER –

electrical installation B.Q. page 3/1 item A

(g) 2. The list of basic prices attached to this Form of Tender must be completed, signed and submitted with the tender.

(h) 3. The tender must be returned, sealed in the special envelope provided which must not bear any mark indicating the name of the tenderer. It is to be sent to the Chief Executive and Clerk of the County Council, Room 10, County Hall, Southill so as to arrive not later than

4.00 pm on 10 October 1982

(i) 4. The Employer is not bound to accept the lowest or any tender.

Fig. 3.5 Cont'd.

It should be noted that the tender figure only is to be stated with no further qualifications indicating that the contractor's price is based on all the conditions, restrictions and obligations applicable, regardless of how fair or unreasonable these may be. The contractor is deemed to have taken all such factors into account when preparing his tender. A qualified tender, i.e. one which assumes certain conditions to apply which are not stated in the tender documents, will almost certainly be rejected or, at best, the contractor may be given the opportunity to withdraw the qualifications and confirm that his price still stands.

In extreme cases where management feels that certain conditions imposed are too onerous to be accepted or where information is sketchy or ambiguous, a covering letter outlining the problem may be attached to the form of tender. The extent of the qualification will then be considered by the client and accepted or rejected at his discretion.

The tender should be submitted in a plain envelope bearing only the delivery address and a description of the proposed contract at most. On no account should the contractor's identity be revealed on the envelope by way of the company's franking machine, etc.

The tender should if at all possible be delivered by hand ensuring its safe arrival and acknowledged by the architect or client preferably with the issue of a receipt. Whereas the postal service is probably quite adequate for most everyday needs, it must be appreciated that the tender not only has to be delivered on a certain day, but also by a specific time on that day. And if, as is usual, the estimator is working up till virtually the last minute, the post cannot be relied upon to deliver the tender on time and a great deal of work could be wasted since tenders submitted after the deadline are usually not considered.

Notes to Fig. 3.5

(a) The tender figure written in words and figures.

(b) Although there is no fixed time limit in which the client must accept or reject a contractor's offer (i.e. the tender), in this case the employer clearly states that the tender is to remain open for acceptance for a period of three months from the date of tender. The estimator must bear this in mind when pricing since a firm price contract of twelve months' duration effectively becomes fifteen months as far as allowing for the effects of inflation is concerned. If the tender were not accepted within the three months from the date of submission, the offer would normally lapse by passage of time and the contractor should be allowed to re-price the work.

(c) In this case, the interests of the clients are to be protected by a bond taken out by the contrator with an insurance company, the name of which must be stated. The contractor confirms that the cost of the policy is included in the tender (see also Chapter 6, 'Preliminaries').

(d) On local authority contracts entered into under the JCT Standard Form Local Authorities edition, the contractor must comply with clause 19A, 'fair wages clause', under which he agrees to observe the minimum rates of wages agreed by the unions. The contractor confirms here that he is operating under the general conditions stated. (Although it is unlikely to occur, failure to comply with this clause results in breach of contract, and gives the employer the right to determine the contract.)

(e) A responsible person authorised to sign on behalf of the contractor enters his signature together with the company's stamp.

(f) Under the terms of the contract, where the contractor, in the normal course of his business, carries out work which is covered by a PC sum in the bill (e.g. electrical or heating work), he is entitled to be given the opportunity to submit a price when at a later date tenders are obtained for such work. The contractor must state here the PC sums for which he desires to tender.

(g) The contractor is required to submit with his tender a completed 'basic price list' (page 164) since the work is to be executed on a fluctuations basis under clause 39.

(h) Details of place, time and date by which the tender must be submitted.

(i) As is the case with virtually all proposed building work, the employer states that he is not bound to accept the lowest or any tender.

The tender having been thus completed and submitted, all the estimator can do is wait with his fingers crossed for the outcome. On occasions, after the submission of tenders, rival estimators will often contact each other to exchange prices and whilst it is admitted that confidential information is being given to the firm's competitors, valuable information is also being gained relating to current market conditions. In many cases this is the only way the estimator has of discovering how well or how badly he has fared in the competition since many clients neglect to follow the recommended procedure of supplying all tenderers with a list of the tender prices.

Where the tender is unsuccessful, the contractor may hear no more unless he takes the initiative and rings the architect or quantity surveyor, or he may receive a short, polite letter stating that '. . . we regret on this occasion that you have been unsuccessful . . .'! When the result of the tender is known, it is important to record all information relating to the pricing of the job, reports, other tender figures if known, etc., for future use.

Sub-contractors who have submitted quotations should be informed of the outcome of their efforts as promptly as possible even if the tender is unsuccessful.

3.3 Procedure following notification of lowest price

On occasions where his tender is the lowest, the estimator is asked to submit a copy of his priced bill of quantities for checking by the quantity surveyor to ensure that the total of the priced bill is the same as the tender figure. The preparation of a fully priced out bill of quantities is likely to take a number of days since prospective sub-contractors will have to be contacted and individual rates obtained where perhaps only lump sum quotations had been submitted earlier. Furthermore, the estimator himself may have provisionally inserted lump sums for work often represented by several bill items, each of which must now be rated separately. His sum for overheads and profit must be incorporated into the bill, either being reduced down into smaller lump sums and included with certain preliminary items or spread evenly throughout the bill by increasing all rates by the appropriate percentage.

Since the production of an accurate tender demands the detailed pricing under pressure of several hundreds of bill items, it is not surprising that mistakes in pricing, arithmetic, carrying page totals forward, etc. often occur and which come to light only when the priced bill is submitted for checking. This means that the priced bill of quantities differs from the original tender figure by a greater or lesser amount depending on the nature of the mistakes which are aggregated to produce a net result. Such mistakes must be rectified before any contract can be entered into and it is the responsibility of the quantity surveyor to examine the bill, identify any errors and inform the contractor accordingly.

The subsequent procedure which must be followed in order to rectify the position can be either one of two methods available as recommended in the Code of Procedure for Single Stage Selective Tendering as follows.

Alternative 1

The essence of this method is that the contractor is not allowed to alter his tender figure. Details of any errors discovered by the quantity surveyor are forwarded to the contractor who, after considering carefully the implications, must either confirm his tender figure or, if the mistakes are too large to absorb, withdraw whence the same procedure will be followed with the second lowest tenderer and so on until the objective is achieved. However, despite errors in his pricing, a contractor may still choose to go ahead with the contract, particularly if he is short of work, in which case his priced bill must be reduced (or increased) by an amount by which the corrected priced bill exceeds (or falls short of) the tender figure, the contractor thereby absorbing the loss (or extra) within his original price. The subsequent adjustments which would be needed where a contractor confirms his tender figure as described are effected by attaching an endorsement to the priced bill indicating that all rates (excluding preliminary items, contingencies, PC and provisional sums) are to be reduced or increased in the same proportion as the corrected total of priced items exceeds or falls short of such items.

Alternative 2

Where alternative 2 is adopted by the client, the contractor is given the opportunity, if he wishes, to alter his tender figure should errors be discovered as described above. However, where a contractor chooses, say, to increase his tender by the amount of any mistakes identified, he takes the risk of not then being the lowest tenderer, in which case his tender would be discarded in favour of the new contractor who would now be the lowest tenderer and the same procedure followed once more. Where the contractor suspects that competition has been fairly close, he can elect not to amend his tender figure if he feels that in doing so he may lose the contract, in which case an endorsement is required as in the case of alternative 1 and the bill rates will be deemed to be adjusted accordingly so that the corrected total of bill items once more matches the tender figure.

Note

It is advisable for a client to decide prior to the preparation of the bill of quantities which alternative is to be used in the event of errors occurring in pricing, and the method to be adopted stated clearly in the bill in order that all tenderers are aware of the procedure to be followed before tenders are submitted thereby avoiding any accusation of unfairness at a later date.

3.4 Overheads and profit

Having outlined the general procedure to be followed in the preparation of the tender, the individual elements which comprise the tender figure can now be examined in greater detail, commencing with what is arguably the principal objective of the contractor, that of securing the recovery of overheads and profit on his building operations.

3.4.1 Overheads

In compiling his preliminary estimate, the estimator has, thus far, based his prices on *net costs* only, ignoring temporarily all other 'invisible' expenses such as the running of the business, etc. Such additional costs over and above the basic cost of carrying out each individual operation described in the bill of quantities are known as 'overheads'.

The prime concern of the estimator when pricing any contract is the accurate assessment and allocation of costs to items and operations which will cover all the contractor's obligations and liabilities, and only then, after satisfying himself on this score, can he turn his attention to the question of profit margins and the extent of the price 'mark-up' he feels he can safely build into his tender figure and yet remain competitive.

For the purposes of tendering, costs may be classified as being either '*Direct*' or '*Indirect*'.

Direct costs

Direct costs are those which can be directly associated with the carrying out of the builder's contractual obligations on site, and can be further sub-divided into those associated with the actual production of the building work such as materials, plant and site labour costs, and those which constitute the expense involved in providing site 'back-up' facilities and services without which the building work itself could not be executed. Costs falling into this latter category would include those items priced in chapter 6, 'Preliminaries', and would include the cost of supervision, temporary electricity supply, water, accommodation, transport and travelling etc.

Indirect (head office) costs

The costs involved in running the site and in producing the actual building itself can be fairly easily identified and allocated within the bill of quantities so as to form part of the overall tender figure. However, in addition to these costs, the contractor incurs much more expense indirectly simply as a result of running his business. Most contractors will be working on a number of contracts at any one time, all at various stages of completion, but none of which could operate independently without the facilities provided by 'Head Office' even though all site charges had been allowed for within each individual tender. These 'Head Office' charges or 'general overheads' can be considerable and will often cover a great variety of activities undertaken by the firm, particularly where an organisation operates a number of different companies within a group since, in addition to running a construction company, a firm may well diversify into plant hire, concrete block manufacturing, precast concrete flooring, haulage, etc., with each company requiring the services of Head Office.

Although every company's policy and costs will vary in detail, the following list is an example of the typical services provided by Head Office and illustrates the extent and variety of the activities performed:

1. *Directors' salaries and expenses, staff salaries.* Probably the largest single item of cost in the contractor's budget and would include the salaries of contract managers, quantity surveyors. engineers, estimators, assistants, bonus surveyor, accountants, clerks and typists, etc.

2. *Receptionist and telephone charges.* Probably including an internal private telephone system.

3. *Stationery, postage, books, etc.*

4. *Company cars.* Whether bought or hired, together with all incidental costs including road tax. insurance, depreciation (if purchased), servicing and repairs, fuel, etc.

5. *Rates.* General rates and sewerage rates will be payable on all buildings owned by the company, the annual bill depending on the value of such properties.

6. *Rent.* Payable to the owner of the property if the accommodation is rented, or mortgage repayments on loans taken out to buy or build offices and other buildings.

7. *Bank charges.* Interest charges payable to the bank on overdrafts and loans.

8. *Electricity and fuel.* For heating, lighting and power in offices and workshops.

9. *Maintenance.* An allowance for repairs to the property, renewal of defective items, repainting, etc.

10. *Insurances.* In addition to labour insurances, which would probably be included in the labour all-in rate or in preliminaries, other incidental insurances which must be allowed for would include the insurance of buildings and contents, work under construction on sites and materials against fire. Although individual policies for each site could be arranged, it is more likely that a 'blanket' policy would be obtained covering the full range of the contractor's building activities throughout the year and a lump sum premium calculated based on total turnover. Where this is the case, it is easier for the cost to be allowed for under general overheads, rather than split and allocated to individual contracts.

11. *Company pension schemes.* The amount of the company's contribution to private pension schemes for employees. (Other costs in connection with the employment of Head Office staff such as national insurance, holidays with pay, etc., would be included under the heading of 'staff salaries'.)

12. *Cleaning.* Employment of cleaners or caretakers, together with the cost of cleaning materials and equipment, or the hire of an industrial cleaning firm.

13. *Replacement of furniture etc.* New filing cabinets, office desk equipment, chairs, etc.

14. *Entertaining, lunches. staff canteen, etc.* Covers the cost of luncheon vouchers, subsidised meals, business lunches, etc.

15. *Computer hire.* Many firms now use computers to handle wages and accounts and, if the equipment is not owned, the company can hire the use of a computer terminal. Other equipment such as photocopier, plan copier, etc., can also be hired from suppliers.

16. *Audit fees.* The annual costs of an independent firm of accountants in auditing the books and authorising the company's accounts.

17. Advertising. Could include draughtsmen's or artists' salaries, consumables and advertisements in the local or national press, trade journals and magazines.

18. Salesmen, cars and commission. May be employed in selling the company's products, visiting architects, prospective clients, etc.

19. Journals, magazines and newspapers. Most firms take these in order to scan the pages for tenders invited and to keep abreast generally with latest developments, new products and prices.

20. Drawing office charges. Staff and equipment where the contractor is involved in speculative house building requiring design work and submission of proposals for building and planning approvals.

21. Interest on working capital, loans, retention money. A typical contract will provide for interim payments to be made to the contractor at regular (usually monthly) intervals throughout the job after the first month's work has been completed. However, despite this provision, the contractor is obliged to finance at least the first month's production himself and where he has a number of contracts all in the early stages of construction, the amount of 'working capital' needed can often only be found by borrowing from the bank. Furthermore, throughout the year there will always be money outstanding owed to the contractor in the form of retention held by clients. During times of high interest charges, the cost of borrowing to finance contracts and the cost of money outstanding can be considerable.

The above list is not exhaustive, but is intended to illustrate the nature and variety of costs involved in simply 'running the business' and such costs should not be underestimated or dismissed lightly.

Table 3.1, based on the assumed overhead costs above, shows the extent of the sums involved for a typical small to medium-sized contractor.

The task of establishing the figures would not normally fall within the estimator's province, but would be assessed by the company accountant and would represent the cost of running the business for a whole year. It is likely that the current year's budget figure would be drawn up at the end of the company's previous financial year and would, in essence, be a prediction of the coming year's expenditure. There exists here a similarity between the accountant's and the estimator's responsibilities with both having to look back at previous costings, expectations and actual achievements, adapting the results and projecting figures forward into the future

Table 3.1

Item		Amount per annum (£)
1	Staff salaries, including directors' fees and cost of employing staff	150,000
2	Receptionist and telephone charges	11,000
3	Stationery, postage, books, etc.	5,500
4	Rates	2,500
5	Rent (or mortgage repayments)	6,000
6	Bank charges: interest on working capital and retention money	9,000
7	Electricity and fuel	7,500
8	Maintenance	5,250
9	Insurances	2,000
10	Company pension schemes	12,000
11	Cleaning costs	4,000
12	Replacement of furniture, etc.	2,500
13	Entertaining, etc.	5,750
14	Computer and equipment hire	3,000
15	Auditors' fees	1,750
16	Advertising costs	10,000
17	Company cars and expenses	25,000
18	Salesmen, commission, etc.	18,000
19	Journals, magazines, newspapers	500
20	Drawing office charges	16,000
	Total	£297,250

based on necessarily arbitrary assumptions of inflation, demand, etc. Throughout the year, records of actual costs would be kept and compared with the budgeted figures: then, by comparing these, the budget can be adjusted for future years.

The accountant then, having prepared his budget, expects the coming year's overheads to amount to nearly £300,000 – a sizeable sum by any standards – and if the company is to survive, this sum of money must be recovered during the next twelve months' trading and this is before we even start to think about profit! This overhead figure of £300,000 cannot be directly allocated to particular contracts or other sales outlets within the company because the cost of providing the Head Office services covers the *whole range* of the contractor's activities and so the cost is *apportioned* to the various 'cost centres' (i.e. contracts, etc.) on some suitable basis.

When business is slack, some of the overhead charges may be reduced and some may even be postponed for a while, such as repayment of loans, etc.; maintenance could be deferred; advertising reduced; bank charges could be less if the actual level of borrowing falls below that anticipated or where interest rates generally come down. However, most of the costs are fixed irrespective of output and cannot be avoided. Examples of these are staff salaries (the most expensive item of all by a long way), rates, fuel and power, auditor's fees, etc.

The usual method of recovering these overhead costs is by levying a percentage addition of all costs incurred in the provision of goods and services to a client. Thus within any tender figure for a contract, there will be included a proportion of overhead charges. The actual overhead percentage to be charged on each 'sale' (or contract) can be determined by making a prediction of the company's likely *turnover* for the coming year in a similar manner to the method used for establishing the overhead figure itself.

For example, assume a typical company involved in other areas of business in addition to contracting; turnover may be anticipated as shown in Table 3.2.

Thus if the previously calculated overhead charges are evenly apportioned between the various 'profit centres' above, the overhead percentage to be charged on all 'sales' will be:

$$\frac{£297,250}{£3,375,000} \times 100\% = 8.81\%$$

This assumes that an allowance of 8.81 per cent for overheads had previously been incorporated into the value of work now classed as 'in progress'. If a smaller percentage had been used, then obviously a further adjustment would have to be made. However, the management may feel that it would be inequitable for the overhead percentage to be applied equally on all the activities; indeed, when work is in short supply the contracting side of the business may be unable to compete if it was required to include such a figure in its tenders, whereas concrete block sales may be buoyant and able to incorporate a much higher overhead percentage in its prices. Further adjustments could therefore be made with each of the company's activities bearing a different percentage of the overheads burden in relation to its ability to meet its targets, providing overall, at the end of the trading year, a minimum sum of £297,250 had been recovered.

The hypothetical calculations shown above, however, are fraught with uncertainties since all the figures are based on assumptions which may or may not be realised. The budgeted target for recovery of overheads will only be achieved:

1 If the predicted cost schedule prepared by the accountant turned out to be close to *actual* costs; and
2 If the company's turnover as a whole reaches the anticipated levels.

Table 3.2

Activity	**Anticipated turnover (£)**
(i) *Building* All work in progress (i.e. the total value of contracts unfinished at the end of the previous financial year)	1,000,000
(ii) *Speculative housing development* Assume 10 completions within the next 12 months at an average cost of £30,000 each	300,000
(iii) *New contracts* Assume 3 new public works contracts will be secured at (say) £500,000 each = £1,500,000, assume full completion of one within 12 months, 50% completion in the second and 20% completion in the third; therefore, turnover is: 100% of £500,000 £500,000 50% of £500,000 250,000 20% of £500,000 100,000 £850,000	850,000
Assume 2 new private industrial/commercial contracts average £400,000 each, one fully complete within 12 months, other one 50% complete; therefore, turnover is: 100% of £400,000 £400,000 50% of £400,000 200,000 £600,000	600,000
Jobbing work, small works, repairs, etc. (say) £100,000	100,000
(iv) *Concrete block sales* (say) £50,000	50,000
(v) *Concrete beam and block flooring*, supplied and fixed on a sub-contract basis (say) £400,000	400,000
(vi) *Income from hired plant, haulage etc.* (say) £75,000	75,000
Total anticipated annual turnover	£3,375,000

Notes:

(a) In addition, assets could be revalued which would in effect boost the turnover further.

(b) The above figures and range of activities have been simplified to illustrate the principle involved. In practice, the accounts would be much more complex.

Where actual costs exceed budgeted costs and/or turnover fails to reach the expected figure, the target will not be reached and the company will *under-recover* its overheads, a position which could not continue for long without serious consequences. Alternatively, where actual costs fall short of budgeted costs as a result of savings made or an unexpected fall in inflation and/or turnover is higher than anticipated, then an *over-recovery* of overheads will be achieved which effectively will mean extra profit for the company.

Having once established a relationship between turnover and overheads, it can be used by management to aid policy-making decisions in the future. For example, there may be indications that overheads next year are expected to increase from the current level of £297,250 by 10 per cent to a new figure of £326,975.

Management must now decide whether turnover can be increased by a corresponding percentage in order to keep pace or whether the overhead percentage itself will have to increased. Where neither of these alternatives are possible, savings must be made in the overheads, probably a difficult task to achieve without making staff redundant since most of the overhead costs are fixed irrespective of turnover.

Thus, assuming a revised overhead target of £326,975 to apply, the following adjustments would have to be made in the light of the economic climate at the time:

Case 1

A situation where the economy is healthy, investment is taking place and the demand for building is high. In such a case, management may reasonably expect the business to expand and turnover to be increased by, say, 13 per cent on average.

Previous year's turnover	£3,375,000
Expected increase – 13% to	£3,813,750
Overhead figure established at	£ 326,000

Therefore overhead percentage to be charged

$$= \frac{£326,975}{£3,813,750} \times 100\%$$

$$= \underline{8.57\%}$$

Where the overhead percentage can be *reduced* as in this case, the company's competitiveness would be further improved, hopefully leading to the securing of even more work, although it must be remembered that when trading conditions are favourable for one firm, they are likely to be so for its competitiors as well who will also have the opportunity of reducing *their* overheads. It is possible, therefore, that the relative advantage of one firm over another will not change.

Case 2

A situation of standstill in the economy with fewer orders. Increased inflation will keep turnover constant, but overheads will be increased as in Case 1.

Previous year's turnover £3,375,000
Expected increase – nil
Overhead figure established at £326,975

Therefore overhead percentage to be charged

$$= \frac{326{,}975}{£3{,}375{,}000} \times 100\%$$
$$= \underline{9.69\%}$$

In this case a greater burden of overheads must be shouldered by the various parts of the company, leading to higher prices being charged, which in turn could lead to fewer orders.

Case 3

During a recession where a serious decline in workload is anticipated, inflation will still push overheads upwards, but turnover can be expected to fall in real terms. The consequences of, say, a 10 per cent decline in turnover would be as follows:

Previous year's turnover £3,375,000
Expected decrease – 10% to £3,037,500
Overhead figure established at £326,975 (as before)

Therefore overhead percentage to be charged

$$= \frac{£326{,}975}{£3{,}037{,}500} \times 100\%$$
$$= \underline{10.77\%}$$

In this case the company is faced with a dual problem of declining demand coupled with increasing overheads – potentially a disastrous situation calling for immediate positive action by management either to increase turnover, perhaps by reducing profit margins or diversifying into a more profitable field, or by reducing overheads in some way.

Forecasts such as the above examples illustrated will no doubt be prepared by the company accountant, perhaps in conjunction with senior management and perhaps the estimator himself depending on the size of the company. Whatever path is chosen, there is no doubt that the prime concern of the company is the recovery of its overheads and failure to do this will very quickly bring about the failure of the company.

3.4.2 Profit

Profit may be defined as being that amount of income which remains after all production and other costs have been deducted. Traditionally in the building industry, profitability has, on average, been very low in comparison with other industries, although from time to time when conditions permit, building can be very profitable indeed.

However, the uncertainty which results from an uneven workload leads to problems in forecasting turnover accurately and an inability to plan ahead for a given rate of return on capital employed. In an industry which is slow to respond to rapid changes in an economic scene which can often move very quickly from recession to expansion and vice versa, it is not surprising that extremes in a contractor's trading position are the rule rather than the exception with carefully laid planned profit margins giving way to a policy of 'charging what the traffic will bear'!

Thus far, the estimate is built up on the basis of net cost with thoughts of required profit levels not yet complicating the issue. Whereas the costs of production, site overheads (in the form of preliminaries) and the assessment of general overheads can all be calculated with a fair degree of accuracy in a purely 'mechanical' way, the same cannot be said of the profit margin. This, the last element to be assessed before the tender is finalised, cannot always be established by the application of a simple predetermined percentage of predicted costs. The amount of profit to be included in the tender is perhaps the most arbitrary item of all and is dependent on a number of different factors all of which must be carefully considered by management in consultation with the estimator who, having priced the job, will be the person at this stage most familiar with the project and who will have developed a 'feel' for the right level of profit which will secure the contract. The budgeting for a *standard* rate of return is therefore not possible due to the vagaries of the industry, and management must be flexible enough and quick to adapt to a new situation which may demand either a near non-existent profit margin or a sensible and sometimes sizeable return depending on whether competition is fierce or circumstances where demand exceeds supply.

Individual contractor's policy with regard to profit is naturally very confidential and each one will no doubt adopt its own particular method of establishing the sum involved. Some may assess their profit as a percentage of *total* costs, including sub-contractors' prices, whilst others may base their return on the value of their *own work* only, being satisfied to include sub-contractors' quotations within the tender at net cost, with no added profit. Alternatively, a straightforward lump sum could be added without regard to a percentage, a method particularly suited to small works involving alterations and repairs where a contract of perhaps £15,000–20,000 would not be worth tackling for a clear profit of less than, say, £4,000, bearing in mind the complications and risk involved. This represents a profit of 20 per cent which, under these circumstances, may be considered reasonable; however, were such a percentage to be added to a £1 m contract it would be unlikely to produce a competitive tender with £200,000 worth of profit included! In general, the greater the value of the contract, the smaller the percentage of profit the contractor would find acceptable.

Whatever method is used to assess the profit required, the earning of that profit will be the contractor's main objective, with the rate of return reflecting to a large degree the company's success and efficiency. Although a company may exist for quite a while gaining no profits at all but merely recovering its costs and overheads, such a situation could not continue indefinitely and a company unable to generate profits would eventually go out of business.

The reasons why profit must be earned can be summarised as follows.

1 Many companies are financed by shareholders who have invested in the business and are looking for a reasonable return on their investment. Whereas in the short term a shareholder may forgo any dividend if the company is in difficulties, in the long term he will be looking for a return in relation to the level of risk and uncertainty attached to his investment. Such returns can only be paid for out of profits.
2 When conditions permit, a company will wish to expand, increase its turnover, diversify, take on extra staff, etc. The money required to embark on such ventures would be financed by previously earned profits if bank borrowing (and therefore interest charges) is to be avoided.
3 Profits may be needed to cushion the effects of a recession. If redundancies and contraction are to be avoided, sufficient reserves must be accumulated during the good times in order to maintain the business intact when the workload declines.
4 A successful and healthy business is founded on the confidence of its employees and creditors and there is nothing more damaging than rumours of cash flow problems, stagnant production or unprofitability. Profits must be earned to maintain confidence and to satisfy both creditors and employees of the continued existence and growth of the company.
5 Where reasonable profits are being made, the company will find it much easier to attract external capital for new projects which it considers are of vital importance to its well-being and which in the future are expected to generate further profits.

No real guidelines can be given in regard to the actual amount of profit to be allowed for in a tender: the figure could range from NIL to 30–40 per cent depending on how badly the work was needed, level of competition, degree of risk and uncertainty, the willingness to work for a particular client and so on. Indeed, profit maximisation may not be the contractor's main aim, with the company policy being directed to other objectives such as gaining an increased share of the market by keen tendering or the establishment of a good reputation for high-quality work at a reasonable cost. Targets such as these if achieved may well secure the firm's existence during a recession whereas a policy of making as much money as possible during good times could alienate prospective clients at a time when the order book is desperately low!

The overheads and profit having thus been determined for the particular contract being priced, the tender summary can be completed as shown earlier. At this stage the lump sum figure will be submitted comprising the total of the *net* rates in the bill, a total sum for all the domestic sub-contractors' quotations, possibly a tender adjustment in respect of firm/fluctuating prices, and a figure for overheads and profit. If the tender appears to be the lowest submitted and the contractor is asked to submit his priced bill for checking, then all these lump sums must then be incorporated into the bill rates as described in section 3.1.

Chapter 4

The cost of resources

4.1 Labour costs

4.1.1 The cost of labour and the National Working Rule Agreement

The terms and conditions of employment for building trade operatives agreed to over the years by tradition and negotiation between employers (contractors) and unions will form the basis of the total labour cost contained in every building contract. It is therefore essential that the estimator possesses a thorough and up-to-date working knowledge of these terms and conditions and can translate the contractors' obligations to his workforce in terms of the cost involved.

Arising out of localised arrangements governing the employment of labour, a standard document setting out the terms and conditions of employment on a national basis has been established, though provision is still made for the retention of regional rules and variations. This standard set of rules is known as the **National Working Rules** (NWR) for the **Building Industry** and is issued by the National Joint Council for the Building Industry, a body representing all parts of the industry and whose main functions are to fix the rates of wages for operatives and to determine their conditions of employment.

Other duties of the Council include:

1 The determination and administration of training schemes.
2 The consideration of any industrial or economic question likely to have a bearing on industrial relations in the industry.
3 The amendment, as and when necessary, of any rule or training scheme within the rules reflecting changes in current practice.

The National Working Rule Agreement (NWRA) is published for the guidance of contractors and operatives alike, and is a reference book containing much detailed information relating, amongst other things, to various payments, allowances and expenses which are applicable under certain conditions of employment. It is not necessary for the estimator to be able to memorise these tables of allowances, etc., which are constantly being updated anyway, providing he knows where to look in the rules for the information he needs. This chapter therefore simply summarises the NWR without exploring the detailed procedure, payments and entitlements.

Composition

There are twenty-seven rules within the agreement, together with explanatory notes, supplementary rates for joiners' shops, etc., a code of health and welfare and details of schemes relating to holiday pay, training, death benefits and safety recommendations.

NWR 1 and 2

These explain the composition of the minimum weekly earnings which currently comprise two payments, namely: (a) The standard basic rate of wages, and (b) the guaranteed minimum bonus (GMB). Rule 1 also lays down the proportions of the normal full-time rate which is payable to apprentices during their period of training and contains special provisions for the payment of watchmen as well as providing for extra payments for trade chargehands and gangers. Operatives who can satisfy the contractor that they are either (a) qualified benders and fixers of bars for reinforced concrete work, are able to read drawings and set out, or (b) qualified tubular scaffolders employed full-time on that operation, are entitled to be paid at the current standard craft rate.

The remainder of Rule 1 deals with annual and public holiday entitlement together with the means of obtaining payment for such periods, the principles of incentive and productivity schemes, and special rules for woodworking establishment operatives.

Rule 1.2 defines 'availability for work' and establishes the requirements for an operative to qualify for the 'guaranteed weekly wage', which is paid irrespective of time lost due to inclement weather and other stoppages beyond the control of the parties (subject to certain conditions). The guarantee is lost, however, if an employee is not available for work within the meaning of the term, in which case he would be entitled only to payment of the appropriate proportion of his guaranteed minimum weekly earnings for half the number of any hours during which, although now 'available' for work, he has been prevented from working by inclement weather or any other cause beyond the control of the parties as stated above.

Also contained in Rule 1 are provisions for temporary lay-off and the procedure concerning the handling of disputes.

NWR 3, 4, 17, 18 – Extra payments

Extra payments over and above the basic plain time rates may be made to operatives on occasions depending on the job the operative has been asked to carry out.

These extra payments are defined under four main headings:

- Rule 17 – Extra payment for discomfort, inconvenience or risk.
- Rule 3 – Extra payment for continuous extra skill or responsibility.
- Rule 4 – Extra payment for intermittent responsibility.
- Rule 18 – Extra payment for tool allowances.

Rule 17 – Discomfort, inconvenience or risk. This covers the following classes of work.

1 *Work at heights* – additional payment is made on a sliding scale depending on the height at which an operative is working on either 'detached' structures or buildings involving 'exposed' work.
2 *Furnace, firebrick work and acid-resisting brickwork* – extra payments are made according to the type of work involved.
3 *General* – exceptional kinds of work lasting for more than 1 hour involving work in water, close contact with dirt, exposure to dust, spray, etc. and work in confined spaces such as tunnels or deep basements.
4 *Operatives using certain kinds of equipment* – such as mechanically-driven drills, picks or spades; cartridge-operated rivet guns; paint spraying machines, etc.

Rule 3 – Continuous extra skill or responsibility. Operatives engaged full-time on certain specified tasks such as 'timberman' or 'banksman', or who are given the responsibility of attending on, operating or driving items of mechanical plant and equipment are entitled to extra payments. The amounts of such payments vary according to the degree of skill or responsibility involved.

Rule 4 – Intermittent responsibility. Additional payments may be made to operatives who are not engaged full-time on particular tasks, but where such work is carried out intermittently, an extra payment is made in accordance with the *actual* number of hours spent on the operation. The specified duties attracting extra payments under this heading comprise simple scaffolding and dry-cleaning stonework by mechanical processes.

Rule 18 – Tool allowances. Additional payments are made to craftsmen for the provision, maintenence and upkeep of a 'full' set of tools. The payment is made on a weekly basis and applies to joiners, bricklayers, plumbers, masons, plasterers, painters and wall and floor tilers, the amount paid varying in accordance with the table contained in the rules.

Further provision is made under Rule 18 for the servicing of mechanical plant, the storage of tools and the loss of clothing through fire.

In all cases under these rules, the NWRA should be consulted in order to determine the precise amounts of extra payments allowed.

NWR 5

This deals with payment arrangements.

NWR 6 – Working hours

Rule 6 deals with normal working hours which at present are 39, comprising 8 hours per day Monday to Thursday and 7 hours on a Friday. Allowances for meal and tea breaks are also stipulated, together with details relating to starting times.

NWR 7 – Overtime

Rule 7 defines overtime and specifies the rates to be paid to operatives for working overtime and during recognised holiday periods.

At present, overtime rates are paid as follows:

Mon.–Fri. First 3 hours of overtime paid at time-and-a-half, thereafter double time until starting time next day.
Sat./Sun. Up to 12 noon Saturday paid at time-and-a-half, thereafter and all day Sunday double time.

(Work done during holidays may be paid for at double time rates throughout, depending on local agreements.)

NWR 8, 9 – Shiftwork and nightwork

These define shift working and provide for enhanced payments to be made where 'double-day shift' and 'three-shift' working is involved.

Where work is carried out at night, operatives are paid an allowance of one-fifth the normal plain time rate provided that at least three consecutive nights are worked.

NWR 10, 11

These deal with the provision of annual and public holidays.

NWR 12, 13 – Conditions of service

Rules 12, 13 define the conditions of service for employees and contain strict procedures for both parties relating to termination of employment and the steps involved in the dismissal of an operative due to bad workmanship or misconduct.

NWR 14, 15 – Travelling and lodging

Providing an operative lives 7 km or more from the site on which he is working, he is entitled to TWO payments for each day he travels to and from the site to his home. The two payments are as follows:

1 Travel allowance. A sum of money, the purpose of which is to compensate the operative for travelling in his own time in order to arrive on site at the normal starting time.
2 Fare allowance. A sum of money which reimburses the operative for the expense involved in travelling to and from the site, a specified amout being paid irrespective of the means of transport chosen by the man.

The sums of money payable to an operative in respect of travelling and fare allowances are calculated on the basis of the distance which he lives from the site, precise rules being stated concerning the method of measuring the distances involved. The allowances are frequently updated and reference should be made to the table in the NWRA in order to establish the current rate of payment.

Where an operative is allocated work on a site in excess of 50 km from his home and does not travel daily, the above rules are superseded by lodging and intermittent travelling allowances.

Where the contractor himself provides the means of transport, the fare allowance is not payable, though the travel allowance is still applicable providing the travelling is done outside the 'normal working day'.

NWR 16 – Payment for absence from work due to sickness or injury

Subject to certain strict conditions, all operatives over 18 years of age who, during their period of employment, are absent from work on account of sickness or injury are paid a small lump sum by the contractor for each day after the first three days of each period of absence.

NWR 19 – Retirement and death benefit cover

Operatives are entitled to be provided with a retirement benefit and death benefit cover, the premiums being paid by the contractor through the payment of a small surcharge on the value of the weekly holiday stamp.

NWR 20, 21

These deal with the training and employment of apprentices.

NWR 22

This deals specifically with scaffolders' wages and duties.

NWR 23, 24

These deal with safety on sites.

NWR 25 – Trade union recognition and procedures

This lays down rules and regulations regarding trade union officers, shop stewards, meetings, etc.

NWR 26 – Register of employers

Employers registered by the National Joint Council for the Building Industry shall undertake, *inter alia*:

1 To establish direct communication between officials of unions adherent to the NJCBI on appropriate matters.
2 To endeavour to ensure that all operatives are directly employed.
3 To inform regional officials of new contracts and details of anticipated labour demand.

NWR 27 – Grievances, disputes or differences

This lays down the procedure to be adopted in the case of disputes and grievances.

In addition to the above twenty-seven rules, further recommendations which could have an effect on cost and therefore of interest to the estimator are stated concerning the following subjects.

1. Incentive schemes and productivity agreements

The object of such schemes should be to:

(a) increase efficiency, thereby keeping the cost of building work at an economic level.
(b) encourage greater productivity, thereby providing an opportunity for increased earnings by increased effort, while maintaining a high standard of workmanship.

The effectiveness of incentive schemes depends on the co-operation between management and operatives to ensure that the organisation of the work is such that realistic targets can be achieved, whilst at the same time a genuine effort is being made by operatives to improve output.

The general principles of an incentive scheme should be as follows:

(a) A target should be issued by the management for each operation and, according to the extent that the performance is better than the target, an additional payment should be made.
(b) Targets should be issued before an operation starts.
(c) Targets are dependent on the saving rate adopted in each scheme and the proportion of the saving which is to be paid out as bonus should be clearly stated.
(d) Targets should be based on 'reasonable' standards of performance.
(e) Schemes should be expressed in simple, precise terms in order that there is no confusion over earnings or the extent of the work involved in the operation.

2. *Industrialisation of building processes*
Contains a statement of principles to be adopted.

3. *Supplementary rules for woodworking factories and shops.*
Lays down rules for rates of wages, extra payments, repetitive process work, working hours, etc.

4. *Code of health and welfare conditions for the building industry*
Contains provisions for:

(a) Standards regarding safety on sites, provision of first-aid equipment and first-aid rooms.
(b) Standards regarding the welfare of operatives such as the provision of washing facilities, shelter and accommodation for clothing and for taking meals, sanitary conveniences, protective clothing and the provision of safe access to all facilities.

5. *National joint training scheme for skilled building operatives*
Contains a summary of the main provisions.

Conclusion

Many of the foregoing rules and recommendations apply to current practice on the site and as such will have no effect on the tender figure as far as the estimator is concerned. However, a good knowledge of the rules is nevertheless important in order for the estimator to appreciate fully the conditions under which the men on site are employed and the background against which they are required to work.

Certain rules, such as those regarding wages, holidays, extra payments and sick pay etc. are of utmost importance to the estimator as they form the basis of the labour all-in rate calculation, the significance of which is illustrated in the next section.

4.1.2 All-in rates for labour

In any form of building work, the labour content will form a substantial proportion of the total cost, a rough rule of thumb guide allocating approximately 40 per cent of the tender figure to this element on an 'average' contract. In order to be able to price the labour content of any building operation, it is necessary for the estimator to answer two basic questions:

1 *How much*?
(i.e. how much is it costing the contractor *per hour* to employ the operative on the site?) and
2 *How long*?
(i.e. how long will it take the operative to complete one 'unit' of the building operation described in the bill of quantities).

The question of how long it will take a man to perform any particular task is probably the most arbitrary aspect of the whole estimating process and is considered in some detail at a later stage (see Chapter 7). However, the subject of this chapter is the calculation of the hourly all-in rate for labour, a task which is fundamental for the estimator as the figure obtained will form the basis of the forecast of total labour costs with an error of a few pence per hour at this stage leading to discrepancies in the order of perhaps thousands of pounds in the tender figure.

The question of cost per hour to the contractor of employing an operative on the site is referred to as the 'all-in' rate for labour, the term being so called because it embodies not only the cost to the contractor of an operative's wages, but also the costs of employing him and can therefore be looked at from these two aspects, namely:

1 *The direct cost*
(i.e. the wages and other payments which are actually paid to the operative) and
2 *The indirect cost*
(i.e. the statutory and other costs incurred automatically as a consequence of employing labour).

1. The direct cost

(a) Wages. The largest single cost to the contractor is the wages he pays to his men, the NWRA setting out the procedure and qualifications for payment. Wages are paid weekly for a normal working week of 39 hours made up of 8 hours Monday to Thursday

and 7 hours on a Friday, with earnings normally being reviewed annually at the end of June. In addition to this annual review, intermediate adjustments to hours worked, rates, bonuses and other payments are often agreed between employers and unions which affect the cost of employing labour. It is vitally important, therefore, that all changes in nationally agreed rates of wages and conditions of employment are carefully noted by the estimator and the all-in rate altered accordingly.

The current minimum weekly earnings comprise two separate elements:

1 The basic weekly rate
2 The guaranteed minimum bonus (GMB).

The *current* rates for each are as follows:

	*Skilled operatives**	*Unskilled operatives*
(a) Basic weekly rate	£ 88.33½	£75.27
(b) GMB	£ 14.04	£11.89½
Total minimum weekly earnings.	£102.37½	£87.16½

(*Rates apply to skilled workers in the building industry with the exception of plumbing, heating and ventilation, electrical and mastic asphalt trades where different wage rates are applicable.)
The guaranteed minimum bonus is not an 'earned' bonus in the sense of earnings made by way of an incentive scheme, but is a fixed weekly payment for 39 hours worked. The purpose of the GMB is to compensate operatives where:

1 A firm does not operate an incentive or productivity scheme; or
2 Where such a scheme is in operation, but where, for some reason beyond the operatives' control such as wet time, plant breakdowns, etc., bonus earnings fall short of the current level of GMB. In this case the GMB acts as a safety net and any low bonus earnings are made up to the minimum figure.

(b) Overtime. Where an operative works more than the standard 8 hours in any weekday Monday – Thursday or the 7 hours on a Friday, or where he is asked to work at any time on a Saturday or Sunday, he is entitled to be paid for each extra hour worked at enhanced rates in accordance with the following rules;

Mon.–Fri.	First 3 hours at time-and-a-half, thereafter at double time until starting time the next morning.
Sat./Sun.	Time worked from starting time Saturday morning until 12 noon at time-and-a-half, thereafter and all day Sunday at double time.

There is no limit to the amount of overtime which may be worked in a week by an operative, and in practice it will vary from one week to another, from one site to another and from one contractor to another depending on the amount of work available, job progress, length of daylight left at the end of the normal working day, etc. Contractors in general will be very careful about ordering overtime since it is expensive from two points of view. Firstly, higher rates have to be paid to the men and, secondly, productivity is likely to suffer since a full 8 hours will have already been worked and operatives will be tired and looking forward to going home!

(Note: Overtime rates are based on the 'basic weekly rate' only with GMB omitted from the calculation)

(c) Tool allowances. Craft operatives are expected to provide their own hand tools with which to carry out their particular trade. However, since such tools are being used on the contractor's behalf, a small sum of money is paid to the operative to assist in their upkeep and replacement, the payments ranging from 55p to £1.06 per week, depending on the partiecular trade involved, for each week the operative is on site (the full list of allowances for each trade is set out in the NWRA).

(d) Other extra payments. The NWRA provides for extra payments in addition to the normal wage rate to be made where an operative is involved in certain specified tasks (see NWRA 3). Although the sums involved are usually quite small, an accurate tender demands that an estimator identifies areas of work where such payments may be applicable and incorporate them into the all-in rate.

Notes:

(a) *Non-productive time*. Providing an operative makes himself available for work each day within the definition of the term as used in the National Working Rules, he is *guaranteed* the normal minimum weekly wage whether or not the contractor has any work for him, or whether he is laid off due to inclement weather, materials not having arrived on site, plant breakdowns, waiting for other trades to finish, etc. It is for this reason that the normal working week is often referred to as the 'guaranteed week'. Bearing this in mind, the estimator must take into account the amount of lost time anticipated on site and include the cost of this in the all-in rate calculation. In practice, the amount of non-productive time actually encountered could vary considerably from one week to another, from one site to another and from one trade to another depending on the time of year during which the work is expected to be carried out and the degree of dependence a particular trade has on weather conditions. Reliable feedback from previous contracts will assist in arriving at a decision regarding exactly how much time to allow in the tender, but can at best be only a guide and an 'average' figure determined.

(b) *GMB*. Where an operative earns more than the minimum amount of guaranteed minimum bonus in a week, the GMB is not paid in addition. The amount of GMB as stated earlier is a fixed weekly sum; however, it is reduced proportionately for each day an operative fails to make himself available for work.

(c) *Travelling*. Travelling and fare allowances may also be paid directly to operatives in accordance with NWR 14 and 16 and can be incorporated into the all-in rate or, more likely, priced in the preliminaries as a lump sum.

2. The indirect cost

(a) National insurance (NI). National insurance contributions are required, by law, to be made by both employers and employees in all forms of employment, the size of the contribution made by an employer, being considerably more that of an employee. However, since an employee's payments do not form part of the contractor's costs, they will not enter into the all-in rate calculations. NI contributions are made by employers to local offices of the Department of Health and Social Security (DHSS), a Government department which, in addition to handling contributions, also deals with the payment of benefits to individuals who are entitled to claim state support by way of unemployment pay, supplementary benefit, etc. Such payments are funded by the contributions which are made and since the level of 'contributions' and 'benefits' are not always in balance, due to changes in the level of unemployment, etc., the rate at which such contributions have to be made is adjusted from time to time by the Chancellor of the Exchequer, which has the effect of either reducing or increasing the cost of employing labour. The estimator must therefore be aware of any changes implemented and the date on which new rates of contribution come into force. Since the estimator, in tendering, is attempting to predict his costs and therefore, amongst other things, his total NI liability, his task is complicated by the fact that the size of the contractor's NI payments will vary from one week to another during the course of the contract in accordance with

(a) the number of men employed; and
(b) the amount of their gross earnings.

These two statistics will be constantly changing, the contractor adjusting the size of his labour force from week to week in line with the amount of work available, and the wages earned varying from week to week depending on the number of hours of overtime worked and the amount of bonuses earned by way of incentive schemes.

The amount of NI payable by the contractor in any week is based upon a percentage – the current rate being 10.45 per cent of the gross wages earned by each operative up to a specified level of earnings – the current figure being £235 per week. However, since wages will vary from one week to another, the estimator can only assume an average wage level and base his calculations on this hypothetical figure, whereas the actual amount of NI payable will either be in excess of, or fall short of, his 'arbitrary' average as each week passes. Overall, of course, it is anticipated that through this 'swings' and 'roundabouts' principle, payments will even themselves out by the end of the contract.

(b) CITB levy. The Construction Industry Training Board (CITB) was established in 1964 with the aim of providing training opportunities for personnel within the industry. The scheme is funded partly through Government grants and partly from contributions made by contractors themselves, the amount being paid by each contractor varying depending on the number of operatives employed in the previous tax year. Contractors are required to contribute £71 and £18 per annum for each skilled and unskilled operative respectively on the firm's payroll, although where firms employ 'labour only' operatives in addition, the payment includes a percentage of the total 'labour only' payments made, again, during the previous tax year, the current rate being 2 per cent.

(c) Sickness. When an operative is absent from work due to illness, he will not receive any wages, although he may qualify for state benefits as mentioned earlier, together with a payment from his employer, the contractor. At present the payment made by the contractor amounts to £7.35 per day for each day the operative is ill after the first three days (a detailed explanation of the scheme can be found in NWRA – rule 16).

Since the contractor is obliged to make this payment when obviously no work is obtained in return, sick pay is a further cost which must be allowed for. Again, as with NI, the problem is knowing exactly how much sick pay to allow for since some men will be off work due to illness for some weeks in any year whilst others may not be absent at all. At the time when the all-in rate is being calculated, the amount of sick pay the contractor will have to pay out will not be known and therefore some arbitrary average period of illness per operative must be chosen, a typical allowance being approximately 5 days per annum per person, the cost to the contractor being 5 days $\times$ £7.35 per day = £36.75 per year.

(d) Redundancy pay. Should the contractor find himself in the position of having to make operatives redundant, he is required to compensate them by way of a lump sum redundancy payment as

stipulated under the Employment Protection (Consolidation) Act 1978, and in anticipation of such an event must make provision to meet this obligation. The size of the payment to which an operative is entitled varies from one man to another depending principally on the current wages earned and the length of service given to his employer, although the contractor can normally recover 41 per cent of any redundancy payment made to an operative in the form of a rebate from the 'Redundancy fund' held by the Department of Employment.

The usual method adopted by a contractor in providing for such payments should they become necessary is to make an allowance within the all-in rate by including a percentage, normally in the region of approximately $1\frac{1}{2}$ per cent of the gross pay. Again, the actual figure chosen is necessarily arbitrary since the contractor cannot know at this stage what his future redundancy pay commitments will be.

(e) Employers' liability and third party insurance. The contractor must take out an insurance to indemnify himself against claims arising out of injury to persons or damage to property as a result of fire, etc., as the form of contract requires. The cost of the premium will vary from one firm to another depending on the degree of risk assessed by his insurance company, with the extent of such risk determined by: (1) the type of work in which the contractor is normally engaged; (2) the company's turnover within the normal trading year; and (3) the firm's recent record in respect of previous claims.

The premium can be included in the preliminaries as a lump sum if desired, but where the cost is to be covered in the all-in rate, a percentage, say, 2 per cent of gross pay, can be included therein.

(f) Holidays with pay. Operatives in the building industry are entitled, at present, to 21 days' annual and 8 days' public holiday per year, during which time they receive their normal basic weekly wage (including GMB). The cost to the contractor of providing paid annual holidays for his operatives is spread throughout the year by his purchase of 'holiday stamps' – one for each of his employees for each full working week the man puts in on site. These stamps are bought from the 'Building and Civil Engineering Holiday Scheme Management Ltd' and affixed to a card held on behalf of each operative by the contractor. As and when an operative takes part of his annual holiday, the card is 'cashed in' and the corresponding amount of money handed over to the contractor with which he pays his men.

The initial price of the holiday stamp is currently £11.45, which includes a small sum of 75p payable by the contractor as a premium to finance the operation of a death benefit scheme (see NWR 10).

However, the cost of these stamps is increased from time to time as both wages and holiday entitlement increases. Public holidays are excluded from the scheme and the contractor must therefore make his own arrangements to meet this obligation.

Summary of costs to be included in an all-in rate

1 The guaranteed minimum weekly earnings comprising:

  (a) the basic weekly rate; and
  (b) the guaranteed minimum bonus.

2 An allowance for overtime (if worked).
3 Any bonus earned in addition to the GMB by way of incentive schemes.
4 Tool allowances (applicable to craftsmen only).
5 Other payments in accordance with NWR 3, 4, 17 and 18.
6 An allowance for non-productive time.
7 Employers' NI contributions.
8 CITB levy.
9 Sick pay.
10 Redundancy pay.
11 Employers' liability and third party insurance.
12 Holidays with pay scheme.

Calculation of the all-in rate for labour – (a) Skilled operative

The following example illustrates a method of building up an all-in rate for labour employed on the site. However, it should be noted that the calculation is only **typical** of a number of alternative methods in common use and is based on the assumptions stated. In practice, contractors' costs will differ and this, together with variations in policy with respect to overtime working, bonuses paid, allowances for lost time, etc., will produce different final figures, even though the initial basic wages apply nationally.

The establishment of the all-in rate can be conveniently divided into two separate calculations involving:

1 The determination of the total number of productive hours worked on average in one year by an operative on site.
2 The cost to the contractor of obtaining that number of productive hours per year.

The total cost, when divided by the number of productive hours obtained, will give the all-in rate per hour.

1. Total number of productive hours worked per year

	Hours
Standard no. of hours per year = 52 weeks × 39 hrs per week	= 2,028

Add

(a) *Overtime*

Assume 1 hr per day is worked, Mon–Thurs, inclusive for approx. 30 weeks of the year, plus Saturday mornings up to 12 noon. Total overtime per week = 4 days × 1 hr plus 4 hrs = 8 hrs/week × 30 weeks = 240 hrs per year. (The amount of overtime worked in practice would vary from week to week depending on the amount of daylight left at the end of the normal working day, extent of workload, etc.) 240

Deduct

(a) *Holidays*

Annual	18 days × 8 hrs per day =	144
	3 days × 7 hrs per day =	21
Public	6 days × 8 hrs per day =	48
	2 days × 7 hrs per day =	14
		227

(The actual allowance for holidays will vary slightly from one year to another depending on the number of days which fall on a Friday.)

(b) *Non-productive time*

Allow, on average, 1 hour of lost time each day excluding Saturday morning, due to bad weather and other stoppages.
No. of days worked on site in one year:
= (52 × 5) − (21 + 8 (holidays) + 5 (sick))
= 260 days − 34 days
= 226 days × 1 hr per day = 226 hrs per year

(c) *Sick periods*

Allow 5 days per year per employee
No. of hours lost per year = 1 week × 39 hrs
= 39

SUMMARY

Standard no. of hours per year plus overtime (2,028 + 240)			= 2,268
Deduct:	(a) Holidays	= 227	
	(b) Non-productive time	= 226	
	(c) Sick period	= 39	492
			1,776

2. *The cost of obtaining 1,776 productive hours per skilled operative per year*

(a) *Basic wages* £

no. of working weeks $= 52 - (4\frac{1}{5} + 1\frac{3}{5} + 1)$ (annual hols. + public hols. + sick)

$= 52 - 6\frac{4}{5}$

$= 45\frac{1}{5} \times$ £$102.37\frac{1}{2}$ per week = 4,627.35

(b) *Overtime*

Overtime rate $= \dfrac{£88.33\frac{1}{2} \times 1\frac{1}{2}}{39} =$ £3.40 per hour

No. of hours of overtime worked in a year = 240

Therefore cost of overtime = 240 hours at £3.40 per hr = 816.00

(c) *Tool allowance*

Allow, say, £1.00 per operative for each week worked on site

Annual cost = 47 weeks × £1.00 = 47.00

(d) *Sick pay*

Allowing 5 days sick period per year

Annual cost = 5 days × £7.35 per day = 36.75

(e) *Public holidays with pay*

Allow for $1\frac{3}{5}$ weeks at £$102.37\frac{1}{2}$ per week (Holidays actually taken in 8 separate, isolated days) = 163.80

Sub-total 1 £5,690.90

the costs determined thus far will establish the amout of the contractor's NI contributions. None of the items which follow, including the annual holidays with pay stamp, are subject to NI.

(f) *National insurance contributions* at 10.45% = 594.70

(g) *CITB levy*

Lump sum allowance of £71.00 per year = 71.00

(h) *Annual holidays with pay*

Weekly cost of stamp, including 75p for death benefit scheme = £11.45 payable for 47 weeks = 538.15

Sub-total 2 £6,894.75

(A further sub-total is obtained at this stage as the remaining allowances are calculated as percentages of this cost.)

(i)	*Redundancy pay* Allow, say, $1\frac{1}{2}\%$	=	103.42
(j)	*Employers' liability.* Allow, say, 2%	=	137.90
	Total cost per year	=	£7,136.07

$$\text{All-in rate} = \frac{\text{Total cost per year}}{\text{no. of productive hrs per year}} = \frac{7{,}136.07}{1{,}776}$$

$$= \text{£}4.02 \text{ per hour}$$

Calculation of the all-in rate for labour – (b) Unskilled operative. The same principles apply when building up the unskilled hourly rate, except that the wages are different and the tool allowance is not applicable. The calculation is as follows:

1 Assume the number of productive hours worked per year is the same as for the skilled operative, i.e. 1,776 hrs.

2 *The cost of obtaining 1,776 productive hours per unskilled operative per year*

			£
(a)	*Basic wages* $45\frac{1}{5}$ weeks × £87.16$\frac{1}{2}$	=	3,939.86
(b)	*Overtime* Overtime rate $= \frac{\text{£}75.27 \times 1\frac{1}{2}}{39}$ = £2.89$\frac{1}{2}$ per hour × 240 hrs	=	694.80
			4,634.66
(c)	*Sick pay* Allowance as for skilled operative: 5 days × £7.35 per day	=	36.75
(d)	*Public holidays with pay* $1\frac{3}{5}$ weeks × £87.16$\frac{1}{2}$ per week	=	139.46
	sub-total 1		£4,810.87
(e)	*NI contributions* at 10.45%	=	502.74
(f)	*CITB levy* Lump sum payment of £18.00 per year	=	18.00
(g)	*Annual holidays with pay* 47 weeks × £11.45 per week (*Note*: The cost of the annual holidays with pay stamp is the same for both skilled and unskilled operatives.)	=	538.15
	sub-total 2	=	5,869.76
(h)	*Redundancy pay* Allow, say, $1\frac{1}{2}\%$		88.05

(i) *Employers' liability*
Allow, say, 2% = 117.40

Total £6,075.21

$$\text{All-in rate} = \frac{\text{Total cost per year}}{\text{no. of productive hrs per year}} = \frac{£6{,}075.21}{1{,}776}$$

$$= £3.42 \text{ per hour}$$

Notes

(a) The above rates which would be used to price the labour element of the unit rates are exclusive of overheads and profit, and represent only the basic net cost per hour of employing a man. The BQ would normally be priced at 'net cost' in the first instance, and the profit and overheads considered at a later stage.

(b) No incentives or bonus earnings have been taken into account in the calculations, but where such schemes are in operation further provision must be made, either by increasing the allowance for wages in the all-in rate calculations, or by increasing the time allowed for an operation to be carried out when assessing operatives' outputs.

(c) No allowance has been made for employers' contributions towards any private pension schemes on behalf of their employees. Where such schemes are in operation, further provision must be made.

(d) No allowance has been made for extra payments which may be applicable under certain conditions of employment (see NWR). Where it is envisaged that such payments will have to be made, earnings can be increased and the all-in rate adjusted accordingly when pricing that part of the work to which extra payments would apply.

(e) The all-in rate calculations are based on the assumption that operatives are engaged full-time on productive work and no allowance has been made for any supervision costs which would arise out of the employment of charge hands and gangers who are entitled to enhanced hourly rates and who would spend at least part of their normal working day supervising or instructing operatives under their charge. These supervisory costs can be allowed for either by including a lump sum in the preliminaries section, or by inclusion in the all-in rates by dividing the non-productive costs involved between the total number of productive hours obtained:

For example, assume 1 foreman joiner who is paid £20 per week above the normal craftsman's rate, and who is responsible for 8 joiners in his section, spends 50 per cent of his time working and 50 per cent of his time in a supervisory capacity:

Foreman joiners' weekly wages = £102.37½ + £20.00
= £122.37½
Non-productive (or supervisory) cost = £122.37½ × 50%
= £61.19

Total no. of productive hours worked per week:
8 joiners × 39 hours = 312
1 foreman joiner × 39 hrs × 50% = 19½

Total 331½

Therefore cost of supervision per productive hour = £61.19 / 331½
= 18½p (before the addition of any 'on costs')

Conclusion

The all-in rates of £4.02 and £3.42 thus established for skilled and unskilled operatives per hour respectively will be used to price the labour element of the unit rates and all other labour content within the bill of quantities such as that which would be needed in the preliminaries and attendances on nominated sub-contractors, for example.

It must be stressed that the all-in rates are subject to frequent changes in respect of alterations to the terms and conditions of the employment of labour, including rates of wages, holiday entitlement, allowances and expenses, etc., together with changes in the rates of statutory contributions, levies and taxes. It is therefore vitally important that the estimator keeps in touch with all the latest updating of information in order that the effects of all these changes can be incorporated into the all-in rate by adjusting the same in readiness for pricing the next contract.

4.1.3 Daywork calculations

Daywork is a method of valuing work carried out under an architect's instruction when it is not practical to use bill rates or any other method of unit rate build up as defined in clause 13 of the JCT Standard Form of Building Contract. The work is thus not 'measured' as such, but reimbursement is made in accordance with the basic or 'prime' cost of the resources used to which are added percentages to cover all the contractors' other incidental costs, overheads, profit, etc.

The resources which the contractor may employ in order to carry out the work are as follows:

1 The time spent on the operation in respect of skilled and/or unskilled labour.
2 The materials (if any) used in connection with the work.
3 The mechanical plant (if any) needed in connection with the work.

In order to determine the extent of the costs involved in any daywork operation, a record must be kept by the contractor of the time spent and materials used in connection with the work on a 'daywork sheet'. This record must be signed by the architect or clerk of works in order to confirm that the details in respect of labour, materials and plant are correct (see Fig. 4.1).

Subsequently, the contractor and the quantity surveyor may decide to use some other method of valuation for the work and the daywork sheet discarded; however, it is likely that the costs recorded will have a considerable influence on the figure which is finally agreed!

As can be seen from the figure, the quantities of the various resources are priced out using firstly **basic rates** to establish the **prime costs** involved, and secondly a **percentage addition** which makes up the difference between the contractor's prime cost and the price he wants to receive for the work. This section deals with these elements namely: 1. The prime costs; and 2. The percentage additions.

Although employers may decide for themselves the basis of any daywork charges which may arise, the following calculations are based on the definition of the prime cost of daywork as published jointly by the RICS and the B.E.C (NFBTE) and which is in common use on most building contracts.

Labour – the prime cost

The prime cost of labour is expressed as an hourly rate, but differs from the earlier 'all-in' rate in that many of the components of cost are omitted at this stage and are to be allowed for in the percentage addition to be calculated by the estimator as shown later. The involvement of the estimator in daywork studies at the tender stage is dictated by virtue of the method of presentation of the sum to be allowed for this work within the bill of quantities. As can be seen from the example on page 107, the contractor is required to insert his percentage additions where indicated and the cost implications are thus incorporated into the tender figure thereby providing the estimator with a great incentive to keep the percentage as low as possible in order to maintain a competitive price.

However, before a percentage addition can be established, the estimator must be familiar with the items which are included in the prime cost as previously defined. The items so included are as follows:

Daywork sheet. SHEET NO. 24

Description of Work Cut openings for extractor units in existing walls and make good (V.O. No 45)

Contract at: Office conversion, Southtown Client: ABD Associates Week Ended: 12.10.84

Name	Trade	Hours Worked							Total	Rate	£ p	Materials used	Rate	£ p	Plant Used	Rate	£ p
		S	M	T	W	T	F	S									
V. Austin	BK		8	3					11	3.34	36.74	1000 commons	£80	80.00			
R. Backhoe	BK		8	3					11	3.34	36.74	6 bags cement	£4	24.00	Compressor		
J. Madon	LAB		4	3					7	2.86	20.02	1 tonne sand	£11	11.00	and drill		
B. Bradford	PL			6					6	3.34	20.04	3 bags u/c plast.	£4	12.00	4 hrs @	.70	2.80
												1 bag finish	£5	5.00			
												20m angle bead	50p	10.00			
											113.54			142.00			2.80
Percentage Additions										114%	129.44	Percentage Additions	20%	28.40	Percentage Additions	40%	1.12
										£	242.98		£	170.40		£	3.92

Architect/C.O.W. 2.C. Brown Total: £ 417.30

Signed Site Agent. [illegible]

Fig. 4.1 Fully priced-out daywork sheet

(a) The guaranteed minimum weekly earnings.
(b) Any differential payments in respect of skill, responsibility, discomfort, inconvenience or risk as stated in NWRA (if applicable).
(c) Employer's NI contributions in respect of minimum earnings.
(d) CITB levy.
(e) Annual holidays with pay scheme (including death benefit scheme) together with payments in respect of public holidays.

The annual cost of the above items may then be calculated as follows:

1. *Craftsman*

(a) Guaranteed minimum weekly earnings = £102.37½ × no. of working weeks per year		
£102.37½ × 47$\frac{4}{5}$	=	£4,893.53
(b) Employer's NI contributions at 10.45%	=	£511.37
(c) CITB levy	=	£71.00
(d) Annual holidays with pay & death benefit scheme = 47 weeks × £11.45 per week	=	£538.15
		£6,014.05

The hourly cost may then be determined by dividing the above figure by the number of '*standard hours*' worked in a year, this being defined as the total number of normal working hours per year – i.e. 39 × 52 weeks minus agreed annual and public holiday periods – i.e. 39 × 5$\frac{4}{5}$weeks:
No. of standard hours = 39 × 46$\frac{1}{5}$weeks = 1,802 hours.

$$\text{Basic hourly rate} = \frac{\text{annual cost}}{\text{standard hours}} = \frac{£6{,}014.05}{1{,}802}$$
$$= £3.34 \text{ per hour.}$$

2. *Labourer*

		£
(a) Guaranteed minimum weekly earnings = £87.16½ × 47$\frac{4}{5}$	=	£4,166.49
(b) Employer's NI contributions at 10.45%	=	£435.40
(c) CITB levy	=	£18.00
(d) Annual holidays with pay etc. as above		£538.15
	=	£5,158.04

$$\text{Hourly basic rate} = \frac{\text{annual cost}}{\text{standard hours}} = \frac{£5{,}158.04}{1{,}802}$$
$$= £2.86 \text{ per hour}$$

These rates of £3.34 and £2.86 per hour for craftsmen and labourers respectively are used to price out the labour section of the daywork sheet initially prior to the application of the percentage addition. The rates are increased from time to time in accordance with increased wages, longer holidays, etc. and may well change during the course of a contract. However, the percentage additions once established are not subject to fluctuation and remain fixed for the duration of the project, and it is possible that a particular percentage addition having once been determined will be used in future tenders without alteration unless there is some change in the definition of the prime cost itself.

Having established the 'starting point' for evaluating the labour content, the estimator must now calculate the full rate at which he wishes to charge the client for labour employed on a daywork basis. The difference between this rate and the hourly base rate will determine the size of the percentage addition to be included in the tender.

Labour – The percentage addition

In the following illustration, hypothetical conditions have been assumed in order to make the calculation realistic. In practice, of course, contractors' circumstances will differ and each estimator will base his own calculations on the conditions prevailing within his particular organisation.

Since the hourly prime cost includes only certain expenses, the percentage addition must incorporate the following incidental costs – some or all of which will be applicable depending on the nature of the daywork operation.

1. Site staff and site supervision.
2. Additional cost of overtime (unless specifically ordered).
3. Time lost due to inclement weather, etc.
4. Bonuses and incentive payments.
5. Subsistence allowance, fares and travelling.
6. Sick pay.
7. Third-party and employers' liability insurance.
8. Redundancy payments.
9. NI contributions over and above that applicable to minimum earnings.
10. Tool allowances.
11. Use of scaffolding.
12. Provision of protective clothing, safety, health and welfare facilities, lighting and storage.
13. Any variation to the basic rates.
14. Head Office charges (overheads).
15. Profit.

1. Site staff and site supervision. The number and type of site staff and level of supervision needed will depend on the size and complexity of the particular contract and will have been decided upon prior to pricing the Preliminaries section. However, it is likely that any Head Office staff engaged on the project would be allowed for in the overheads, and on this assumption the cost of supervisory staff will be limited to the foreman, trades foremen and gangers. The problem for the estimator is that at this stage the nature of any subsequent daywork operations is unknown and the actual amount of supervision needed could vary considerably depending on the complexity of the work.

Assume average workforce comprises:

18 skilled and 12 unskilled operatives,
1 foreman employed full-time on supervision,
2 trades foremen employed half-time on supervision,
1 ganger employed one-third time on supervision.

Assuming the foreman is paid a salary of £150 per week, the trades foremen £20 per week above the skilled operative rate, and the ganger £10 per week above the unskilled operative rate, the weekly cost of supervision allowing for the time being worked in a non-supervisory capacity would be:

Foreman: £150 × 100%	=	£150.00
Trades foremen: (£102.37½ + £20.00) × 50% × 2	=	£122.38
Ganger: (£87.16½ + £10.00) × 33⅓%	=	£32.39
		£304.77

This cost must now be divided by the number of hours worked on site per week in order to obtain the cost of supervision per productive hour.

The number of **productive** hours worked on site per week would be:

Operatives: 30 men × 39 hrs	=	1,170 hrs
Trades foremen: 2 men × 39 hrs × 50%	=	39
Ganger: 1 man × 39 hrs × 66⅔%	=	26
		1,235 hrs

Therefore the cost of supervision per hour

$$= \frac{£304.77}{1,235} = 25p \ldots \text{to summary}$$

(*Note*: Trades foremen and gangers' time may be included on the daywork sheet providing this is limited to time spent in productive operations only. The respective basic hourly rates for skilled and unskilled labour only may be used in pricing the hours of such supervisory staff with any additional payments in respect of responsibility being covered in the sum of 25p above.)

2. *Additional cost of overtime.* The prime cost does not allow for

any additional costs arising out of overtime worked in connection with daywork and if the contractor envisages the possibility of this, an allowance must be made using the same method as that adopted in the all-in rate calculation earlier. It is possible, however, that the architect may specifically order daywork operations to be carried out at a weekend. In this case the extra cost of the overtime working would be paid to the contractor, though not at the prime cost rate of £3.34 or £2.86. It should be remembered that the GMB element is omitted from overtime rate calculations, any overtime hours being priced at the reduced prime cost hourly rate, making due allowance in the number of hours, for time and a half or double time.

Assuming 9 hours of overtime per week on average are worked, in accordance with the all-in rate build-up, the extra cost would be:

(a) Normal hourly rate $= \frac{£102.37\frac{1}{2}}{39} = £2.63$

(b) Hourly rate allowing for 9 hrs overtime per week

$= \frac{£102.37\frac{1}{2} + (9 \times £3.40)}{39 + 9 \text{ hrs}}$

$= £2.77$

£3.40 being the overtime rate –

i.e. $\frac{£88.33\frac{1}{2}}{39} = £2.26\frac{1}{2} \times$ time and a half

$= £3.40$

Therefore, additional cost of overtime per productive hour $= £2.77 - £2.63$

$= 14$p . . . to summary

3. *Time lost due to inclement weather.* Again, assuming the same conditions as for the all-in rate calculations, i.e. 1 hr per day lost Mon. – Fri.

Weekly cost of lost time would be: 5 hrs × £2.63 = £13.15

which expressed as a cost per productive hour would be $\frac{£13.15}{39}$

$= 34$p . . . to summary

4. *Bonuses and incentive payments (over and above GMB).* Assume a lump sum weekly bonus of £10 per man is paid in recognition perhaps of the fact that daywork operations often involve working in difficult and dirty conditions often with little prospect of an 'earned' bonus being made.

The cost of this lump sum bonus per hour would be:

$\frac{£10.00}{39} = 26$p . . . to summary

5. *Subsistence allowance, fares and travelling.* An explanation of the NWR method of payment to operatives for travelling can be found on page 81, and in order to establish an approximate cost here, an assumption must be made in respect of an average distance a typical operative might live from the site. Obviously this will give only a rough and ready figure since it cannot be known at this stage which operatives will be engaged on daywork activities.

Assume the average operative lives 12 km from the site.

Daily payment = fare allowance of 68p + travelling allowance of (12–6 km) × 7p = 42p
Total daily allowance = £1.10
Number of days travelling per week = 6 (Mon.–Fri. + Sat. morn.)
Weekly payment = £1.10 × 6 = £6.60
And the cost per working hour would be:
£6.60/39 = 17p . . . to summary

6. *Sick pay.* An allowance for possible payments in respect of sick pay would be made as shown in the all-in rate build-up.

Annual cost of sick pay = £36.75 (see earlier)

Cost per hour therefore would be $\frac{£36.75}{48 \text{ wks} \times 39 \text{ hrs}}$

= 2p . . . to summary

7. *Third-party and employers' liability insurance.* Assume the same figure as that used in the all-in rate calculation.

2% of £3.34 = 7p . . . to summary

8. *Redundancy payments.* Again, assume the same figure as that used in the all-in rate calculation.

1½% of £3.34 = (say) 5p . . . to summary

9. *National insurance over and above that applicable to minimum earnings.* NI contributions will be made by the contractor in respect of all earnings by an operative up to the current ceiling. An element of NI payments related to minimum weekly earnings of £102.38 are incorporated into the hourly basic rate as shown; however, any additional NI liability arising out of earnings in excess of this minimum amount must be allowed for in the percentage addition.

Additional earnings assumed:

(a) 9 hrs overtime earnings amounting to (9 × £3.40) = £30.60
(b) A lump sum bonus of £10 = £10.00

Total extra payments per week = £40.60

Additional NI = 10.45% of £40.60 = £4.24

Cost per hour = $\frac{£4.24}{39}$ = 11p . . . to summary

10. Tool allowances. Allow, say, £1.00 per week for tool money in accordance with the NWRA.

Cost per hour = $\frac{£1.00}{39}$ = (say) 3p . . . to summary

11. Use of scaffolding. Any labour involved in the movement, erection or dismantling of any type of scaffolding used in conjunction with the daywork operation will be included in the total number of hours spent on the work as a whole and as such the contractor will be fully reimbursed. However, with regard to the scaffolding itself, a distinction is made between the use of standing scaffolding, i.e. scaffolding already employed on other parts of the contract, and scaffolding which must be brought to site specifically in order to carry out the daywork operation. In the former case, the additional hire cost (if any) must be incorporated into the percentage addition, whilst in the second case, all charges involving hiring, transport, etc. would be recoverable as resources used under the 'mechanical plant' section of the daywork sheet. In many instances, of course, scaffolding would not be needed in connection with daywork activities, but the percentage addition being fixed must take into account all possible conditions.

Assume additional hire costs of using standing scaffolding in connection with daywork amount to, say, £10 per week on average.

Cost per hour = $\frac{£10.00}{39}$ = 26p . . . to summary

12. Provision of protective clothing, safety, health and welfare facilities, lighting, storage, etc. It is likely that advantage would be taken of such facilities already being provided on the site for the purpose of the main contract, and any additional costs arising out of daywork operations would be unlikely in most cases, although the need for lighting and protective clothing is sometimes associated with daywork.

Assume costs relating to additional lighting and protective clothing for use on daywork amount on average to, say, £5 per week.

Cost per hour = $\frac{£5.00}{39}$ = 13p . . . to summary

13. Any variation to basic rates. This applies to contracts where a specified schedule of basic plant charges is to apply for the pricing of the mechanical plant section, the contractor being given the opportunity to allow for updating the stated prices under this heading. Not applicable to the labour section – see mechanical and non-mechanical plant, page 106.

14 Head Office charges (overheads). The allowance for overheads would probably be in line with the percentage used in the main contract, though there may be an argument for the figure to be increased to take into account the element of disruption and extra risk often associated with daywork operations.

Assume a figure of 15 per cent for Head Office charges, this percentage being levied on *total* costs incurred up to this point and would therefore be applied to the foregoing items (1)–(13) inclusive, in addition to the £3.34 basic hourly rate.

(a)	Items (1)–(13) above	£1.83
(b)	Basic hourly rate (prime cost)	£3.34
		£5.17

Add 15% overheads = 78p . . . to summary

15. Profit. Again, the profit required on daywork operations could well be higher than that included in the main contract for the reasons stated in item (14) above.

Assume a profit margin of 20 per cent is required. (As with the overheads, it is likely that the profit element will be calculated on the basis of total costs with overheads being classed as a cost.)

(a)	Items (1)–(13) above	£1.83
(b)	Basic hourly rate	3.34
(c)	Head Office charges	0.78
		£5.95

Add 20% profit = £1.19 . . . to summary

Summary

	Description	*Hourly cost* (£)
1	Site staff and supervision	0.25
2	Additional cost of overtime	0.14
3	Time lost due to inclement weather, etc.	0.34
4	Bonuses and incentive payments	0.26
5	Fares and travelling costs	0.17
6	Sick pay	0.02
7	Third-party and employers' liability insurance	0.07
8	Redundancy payments	0.05
9	Additional NI contributions	0.11
10	Tool allowances	0.03
11	Use of scaffolding	0.26
12	Provision of protective clothing, etc.	0.13
13	Any variation to basic rates	–
14	Head Office charges (overheads)	0.78
15	Profit	1.19
	Total =	£3.80

In order, therefore, for the contractor to be fully reimbursed, his skilled labour would have to be charged out at a total rate of £3.34 (prime cost) + £3.80 (on-costs) = £7.14 per hour.

Thus, the percentage addition which would be needed on top of the hourly prime cost of £3.34

$$= \frac{£7.14 - £3.34}{£3.34} \times 100\%$$

$$= 114\%$$

(*Note*: One percentage only may be inserted in the tender which must cover 'labour' as a whole, i.e. including unskilled operatives, and since a slightly different percentage would emerge if the above calculation were to be based on unskilled rates of pay and allowances, an *average* of the two figures would be used.)

Materials – the prime cost

The prime cost of any materials used in connection with the daywork operation comprises:

1 The invoice cost of the materials (excluding VAT) after the deduction of all cash discounts in excess of 5 per cent.
2. The cost of delivery to site.

Notes:

(a) Where sundry materials of small value are used in connection with daywork, the contractor often supplies these himself out of stocks held in his yard and as such will probably be unable to produce an invoice to substantiate the cost. In these circumstances, the prime cost is based upon the current market value together with any associated handling charges, irrespective of what the contractor actually paid for the materials.
(b) The cost of unloading, storing and handling of any materials used will be included in the total number of hours as stated in the labour and/or mechanical plant section of the daywork sheet.

Materials – the percentage addition

The percentage addition must allow for all those incidental costs as listed for labour earlier under items (1)–(15) which the estimator considers to be applicable, though in practice an allowance for only overheads and profit would be required since any extra charge for small quantities, say, would be included in the prime cost and any cutting waste etc. covered by the total quantity as stated on the daywork sheet.

An average percentage addition assuming only overheads and profit are required may be in the region of 15–25 per cent.

Mechanical plant – the prime cost

This, the third element of the daywork operation, provides for the contractor to be reimbursed for his use of mechanical and non-mechanical plant (excluding non-mechanical hand tools and scaffolding – see percentage addition for labour, p. 99). In this case there are two methods available for valuing the plant, namely:

1 The use of the RICS 'Schedule of Basic Plant Charges'.
2 The adoption of the *actual* rates which the contractor has had to pay for the item of plant.

Either of these two methods will establish the prime cost, to which will be applied, as before, the corresponding percentage addition.

1. The use of the RICS Schedule of basic plant charges. The Royal Institution of Chartered Surveyors publishes a list of both mechanical and non-mechanical plant showing the hourly hire rate for various items of equipment. The list, being very comprehensive and containing prices for the majority of plant items encountered on most building contracts, is updated from time to time in line with increases in plant charges generally, the contractor using the schedule to value the amount of plant used where such plant is already in use on the site.

2. The use of current rates. Where the item of plant needed is not already on the site and must be hired *specifically* for the daywork operation, the contractor is entitled to be reimbursed at the rates he himself has had to pay irrespective of whether that item of plant appears in the schedule or not.

Mechanical plant – the percentage addition

A clause within the Preliminaries section of the bill of quantities will indicate which edition of the schedule of basic plant charges is to be used should the need arise. The estimator can then allow for updating the stated rates in his percentage addition in order to bring them into line with current prices (see item (13) on page 103).

In addition to the above, the estimator must also allow for any other relevant cost listed earlier under items (1)–(15) as before. Items which could be applicable are: time lost due to inclement weather, etc., since the rates are applied only to the time during which the plant is *actually* engaged on daywork: additional cost of overtime; overheads and profit.

An average percentage addition on plant to cover all the above on-costs may be in the region of 35–50 per cent.

Notes:

(a) Where an item of plant is used which is not listed in the schedule, the value may be determined using rates 'reasonably related' to those so included.

(b) The rates in the schedule include for the cost of fuel, oil, grease, spares, consumable stores, licences and insurances where applicable, but exclude the cost of any driver or attendant, whose time will be included in the 'labour' section of the daywork sheet.

4.1.4 Method of incorporating the daywork provision in the tender

The extent of any subsequent daywork operations will not of course be known at the time of the preparation of the bills of quantities, and therefore a provisional sum must be included either in the 'Preliminaries' or 'PC and provisional sums' section to cover the eventual cost, with the sum being adjusted in the final account in the normal way once the total cost of daywork is known.

The most favourable method of incorporating the daywork element into the contract as far as the client is concerned is one which incorporates both the provisional sum *and* the estimator's chosen percentage addition into the tender figure in some way. This encourages him to keep the the figures as low as possible with the result that any subsequent daywork costs will not be as high as they might otherwise be! The method illustrated below (showing a typical method of presentation in the bill of quantities) indicates the effect on the tender figure of the estimator's percentage additions – the higher the percentage additions, the less competitive his tender becomes.

DAYWORK CARRIED OUT PRIOR TO PRACTICAL COMPLETION

		£	p
(A)	Include the provisional sum of £3,000.00 for labour employed on daywork	3,000	00
(B)	Add percentage addition 114%	3,420	00
(C)	Include the provisional sum of £1,000.00 for materials used in connection with daywork	1,000	00
(D)	Add percentage addition 20%	200	00
(E)	Include the provisional sum of £600.00 for used in connection with daywork	600	00
(F)	Add percentage addition 40%	240	00
	To collection	£8,460	00

A further section of daywork covering work which may be carried out during the defects liability period would follow in the bill of quantities, giving the contractor an opportunity to insert different percentages if he wishes. It is likely that in this instance the estimator will require a *higher* percentage addition for each element since at that time the contractor will have departed from the site and the facilities provided in the way of site accommodation, supervision, scaffolding, transport, etc. during the course of the contract will no longer be available and must be provided at extra cost if required.

Conclusion

The examples illustrated in the build-up of the percentage additions are only *typical* of those which may be adopted by a contractor and, as such, are intended to be simply a guide, with the actual percentages being adjusted from one contract to another in the light of the particular circumstances prevailing in that area at that time.

The methods and definitions used in this chapter refer to daywork which is carried out under, and is incidental to, a building contract; where work involves solely maintenance or is of a 'jobbing' nature, other definitions of prime and incidental costs are applicable.

4.2 The material cost

The cost of providing all the building materials necessary with which to complete the works is likely on any type of contract to be considerable, amounting to perhaps 50–60 per cent of the tender figure. It is therefore essential that the contractor's outlay on material purchases does not exceed the total sum allowed for when pricing the bill of quantities, bearing in mind that payment is made in accordance with the rates and quantities of items stated in the bill rather than the actual quantities purchased. Failure to take due account of this will lead to the contractor using more materials than receiving payment for and having to suffer the consequential loss.

The estimator must have, therefore, a sound appreciation of the procedure involving the purchase, control and fixing of all materials used and an understanding of the problems and possible losses associated with each stage of the process from initial ordering to final fixing in position, incorporating all likely costs into the unit rates.

4.2.1 The material element of the unit rate

Provision is made in the bill of quantities for including the cost of materials required on a contract within the rates for individual bill descriptions in each work section, thereby allowing the cost of carrying out each operation, including the material content, to be expressed in terms of the accepted unit of measurement in each case.

However, it should be noted that whilst the cost of materials forming part of the finished work is allowed for in this way, other materials also required albeit on a 'temporary' basis, such as hardcore for access roads and hardstanding areas, chain link fencing and boarding for site security, concrete for ramps and bunkers,

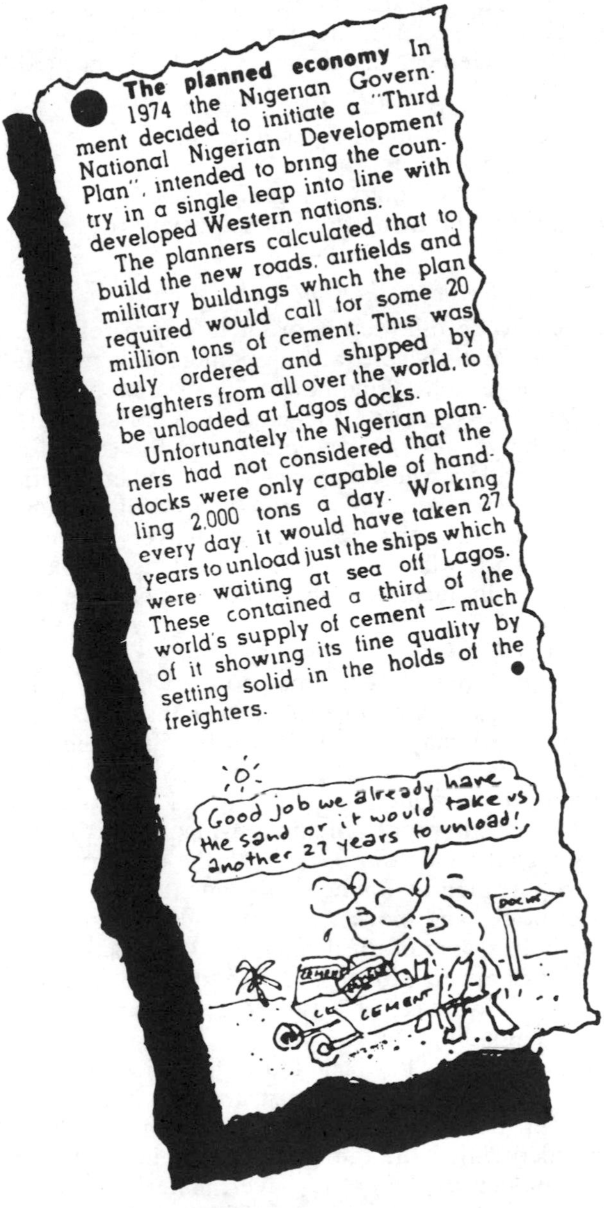

The planned economy In 1974 the Nigerian Government decided to initiate a "Third National Nigerian Development Plan", intended to bring the country in a single leap into line with developed Western nations.

The planners calculated that to build the new roads, airfields and military buildings which the plan required would call for some 20 million tons of cement. This was duly ordered and shipped by freighters from all over the world, to be unloaded at Lagos docks.

Unfortunately the Nigerian planners had not considered that the docks were only capable of handling 2,000 tons a day. Working every day it would have taken 27 years to unload just the ships which were waiting at sea off Lagos. These contained a third of the world's supply of cement — much of it showing its fine quality by setting solid in the holds of the freighters.

Fig. 4.2 Don't forget the incidentals: this newspaper cutting, taken from an article in a recent copy of the *Sunday Express* colour magazine, provides an illustration (albeit extreme!) of the consequences of overlooking the seemingly incidental costs of unloading, handling and distributing materials delivered to 'site'!

cables, pipework and fittings for electricity and water supplies, etc., would normally be priced in the 'Preliminaries' section.

The material element of any unit rate should include for:

1 The price delivered to the site.
2 An unloading, stacking and handling cost.
3 An allowance for waste.
4 The cost of any subsidiary fixing materials.
5 Allowances for the method of measurement of the bill item.

1. The price delivered to the site

The procedure to be followed in order to secure reliable suppliers' prices is explained elsewhere with quotations being carefully examined as suggested, taking note of fixed price periods, cash discounts, whether delivery is included in the price, discounts/surcharges for specified quantities, the cost of returning packing cases, pallets where applicable, etc. Where a price is quoted ex-works or similar, the estimator must allow for the additional cost of providing his firm's own transport for delivery of the materials to the site, although it will be found that suppliers and builders' merchants generally include in their price for haulage to site.

2. Unloading, stacking and handling

Again, quotations must be examined to see whether unloading costs have been included in the price since some materials may simply be tipped whilst others may need careful hand off-loading or even mechanical off-loading. For example, most types of bricks are delivered in neatly strapped packs of around 400 sometimes protected with shrink-proof polythene wrappings; concrete blocks too are often supplied banded in packs with the delivery lorry being equipped with a small hydraulic crane mounted on the back enabling the packs of bricks and blocks to be unloaded quickly and easily with the minimum amount of damage, and whilst there is often an extra charge associated with this service, it would probably work out cheaper to take advantage of it rather than resorting to the tedious task of hand unloading.

Where unloading is not included in the price, account must be taken of the labour and/or plant cost involved in performing this task. A fork-lift truck or crane, for example, may be priced for and its cost included within the rate at the beginning of the work section or in the preliminaries. Often, however, materials have to be hand off-loaded by site operatives which is an expensive method. The pricing of this activity is made considerably more difficult for the estimator by the uncertainty of how long such unloading operations might take, the actual time taken being dependent on the quantity to be unloaded, the weather conditions at the time, the location of designated storage areas in relation to the point of unloading, etc.

Care must be taken by the estimator during his earlier visit to the site, making a note of the site conditions with a view to possible site layouts, ensuring that storage areas are located as close as practicable to access roads and hardstanding areas.

The need for at least some labour to be assigned to unloading operations is unavoidable even when mechanical off-loading is employed since labour in attendance would be needed to assist in manoeuvring packs and crates into stacking or storage positions.

The subsequent handling of stored materials should be reduced to a minimum and confined wherever possible to the actual fixing of the material into its final position in the works, with any intermediate double handling being not only wasteful in terms of labour and possibly plant costs, but also in the increased risk of damage each time the material is moved about.

A further consideration for the need for adequate storage and protection of unfixed materials on site arises from the contractor's contractual obligation to provide these facilities before the value of such items can be included in an interim certificate. However, the cost of storing goods on site can be considerable, particularly if the contractor chooses to purchase some materials at an early stage for reasons perhaps of anticipated delivery problems, and careful thought must be given to the consequences of buying goods early which may have to be stored for several months before being incorporated into the works, for not only is the cost of storage greater, but there is the possibility of further expense where the immediate inclusion of the value of such materials in the next valuation is refused on the grounds of 'premature delivery to site'.

The extent of the cost of providing storage accommodation would vary from one site to another depending on the type and quantity of materials to be used on that particular contract – for example, aggregates, hardcore and sand can be left in the open whereas it would be advisable to cover such items as bricks, blocks, timber, drainage goods, reinforcement, etc. with tarpaulins or polythene sheeting to provide some degree of protection from the weather. Other materials such as cement, plaster, ironmongery, paint, plumbing goods, second fix woodwork items, etc. demand lockable, dry huts for safe storage, whilst for bulky, fragile or very expensive items such as kitchen units, doors, windows, frames and fittings, the contractor may well choose to keep these at his own yard or depot until they are actually needed on site.

Having thus examined the extent to which site storage will be required, the estimator can then calculate the cost of providing the necessary facilities, allowing for storage huts and compounds in the preliminaries and for the protection of completed work within the work sections themselves.

Where nominated sub-contractors are involved in the works, the provision of storage facilities for their materials and equipment is often made the responsibility of the main contractor by the inclusion in the bill of an appropriate special attendance item following the corresponding PC sum for the specialist's work. Where this is the case, the number, type and size of accommodation should be stated, enabling the estimator to identify more readily the cost involved, although the assessment of the length of time such accommodation would be needed on site remains the responsibility of the estimator who would be able to extract this information from the pre-contract bar chart where the time allowed for each nominated sub-contractor to complete his portion of the work would be clearly shown.

The storage of domestic sub-contractors' materials does not normally involve the main contractor in any cost, with the latter usually restricting himself to allocating each sub-contractor storage space on site to erect their own huts if needed.

3. Waste

The wastage of building materials on sites remains a constant and considerable problem, representing a cost running into millions of pounds each year of the nation's energy and raw material resources.

Waste, to the estimator, stems not only from a physical loss of materials due to breakages and damage, but also from any stage of the building process where the contractor is involved in a financial loss by failing to recover in payments under the terms of the contract the actual cost to him of providing those materials. In recognising this unavoidable aspect of material cost, the estimator must allow within his unit rates for a proportion of each material which for one reason or another will be purchased by the contractor but not incorporated into the works as intended. However, whilst the principle may be easy enough to accept, the biggest problem for the estimator is to decide exactly how much waste will be encountered on every building operation on the contract and, given the arbitrary nature of forecasting such events, it is hardly surprising that his original prediction of waste levels rarely coincide with the actual amounts of waste which occur. Although historical records of waste levels may be consulted (if indeed such records exist), the factors which govern the extent of wastage will be different for each contract and can therefore be regarded as a guide only.

Figure 4.3, based on the results of a survey carried out by the Building Research Establishment in 1980, provides an indication of the extent of the discrepancy between estimators' anticipation of waste levels and actual amounts recorded, showing how a contractor can lose out. However, this should not be taken as implying that the estimator does not realise the full extent of the problem. It must be remembered that his principal objective is to be competitive, and the inclusion of extravagant waste allowances in his tender serves only to reduce the competitiveness and consequently the chances of him producing the lowest price. It is a very fine

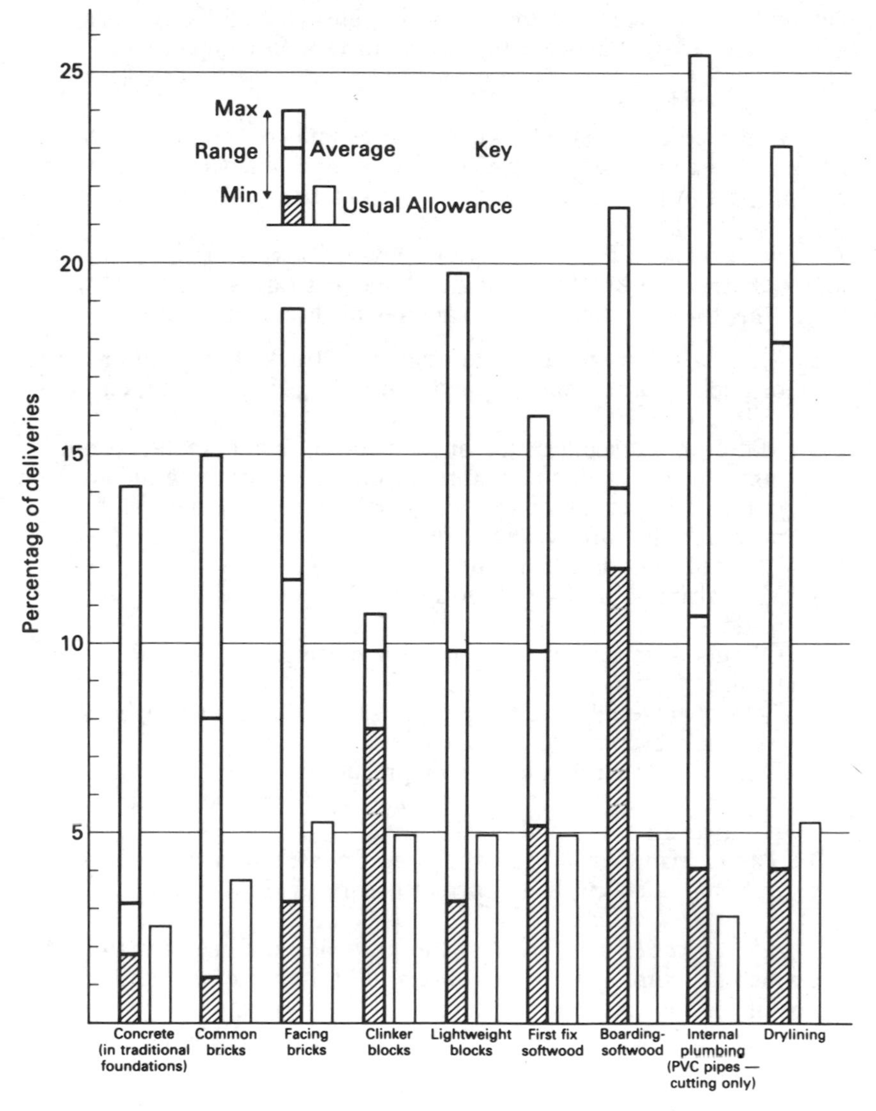

Fig. 4.3 Incidence of waste encountered on building sites compared with estimators' allowances (principal results of direct waste from BRE studies on 230 sites, 1980)

dividing line between, on the one hand, ensuring all likely costs are adequately covered and, on the other, to remain competitive.

The nature of waste

Having established that waste from the estimator's point of view is any aspect of material cost which represents a 'financial loss' to the contractor, waste can be seen as being: (1) direct; and (2) indirect.

1. Direct waste. This involves the physical loss of purchased materials through breakages, theft, items lost on site, rejected work which has to be re-done, etc. Examples of direct waste are:

(a) Materials damaged prior to delivery – bricks towards the centre of packs crushed and chipped, carcassing timbers warped or split, etc.
(b) Materials damaged or lost in unloading, handling and storing operations – inadequate storage conditions resulting in sand and gravel being trampled into the earth, bags of cement and plaster 'going off' in damp huts, etc.
(c) Theft of materials from site.
(d) Vandalism or bad workmanship resulting in work having to be re-done.
(e) Offcuts of materials left over after cutting boards, timber, etc. to size.
(f) Loss of materials due to careless use in fixing – spillages, materials dropped and broken, etc.
(g) Damage caused by other trades resulting in work having to be re-done – feet of scaffolding damaging floor screeds, finished paintwork scuffed and marked, etc.
(h) The carrying out of operations not measured in the bill – the dubbing out of walls to correct bad brickwork, etc.

2. Indirect waste. This involves the purchase and use of materials for which the contractor fails to obtain full reimbursement under the terms of the contract, and whilst there is no physical loss of materials, the contractor nevertheless suffers a financial loss due to the reasons given below:

(a) Broken bricks being used as hardcore.
(b) Damaged facings used as commons where work is subsequently covered.
(c) Materials wrongly specified, resulting in the purchase of better quality materials than required.
(d) The over-ordering of materials resulting from careless preparation of quantities or from ordering directly from the bill where incorrect quantities may be stated.
(e) Surplus materials left over at the end of particular operations – bricks, blocks, mixed mortar, plaster, paint, etc., and whilst such surplus materials may be salvaged for use on future

contracts, the cost of collection and removing to store must be taken into account.

(f) Greater quantity of materials delivered and subsequently charged for compared with amounts ordered – delivery tickets should be carefully checked against orders and invoices.

(g) Higher cost of buying in small quantities – part loads of ready-mixed concrete, cement and plaster bought in odd bags instead of in bulk, etc.

The difficulty for the estimator in trying to determine the extent of the waste involved lies in identifying those stages at which it is most likely to occur since, as can be seen from table 4.1, the wastage of materials, together with corresponding wastage of labour and sometimes plant, generally accumulates at various stages before the work is finally completed and handed over.

In most cases it would be possible to reduce waste or even eliminate it completely if corrective measures were to be taken, but it should be remembered that if the cost of implementing such measures in terms of increased supervision, costly storage and elaborate protection, etc. is greater than the value of the materials saved, then the whole excercise will be pointless and wasted materials will continue to be a common feature on building sites.

4. The cost of subsidiary fixing materials

Whilst the principal material to be used in an operation can be readily found in the bill description, in many cases other 'fixings' are required in order to secure the material in position. Such fixings may or may not be included in the bill description as dictated by the rules of the standard method; however, the estimator must still be able to identify the nature of these fixings, their cost, the waste factor involved and how many are needed per unit of 'work' as measured in the bill. There is a danger when in a hurry to overlook cost of fixing materials, particularly when a material is described in the bill as being fixed 'in accordance with manufacturer's instructions' which, when investigated, may specify the use of some expensive bolts, adhesive or clips, etc.

Examples of subsidiary fixing materials to be incorporated into bill rates would be:

- *Nails* – for fixing timber, tiles, plasterboard, roof decking, etc.
- *Screws, including brass cups and screws* – for fixing joinery work, particularly hardwood.
- *Adhesives* – for fixing floor and wall tiles, certain types of wall coverings.
- *Bolts* – for fixing structural timbers, steel and metalwork, etc.
- *Brackets, clips and connectors* – for fixing plumbing items.
- *Mortar* – for bedding and jointing ridge tiles, jointing precast concrete manhole sections, etc.

Table 4.1 Incidence and types of waste encountered in the construction process

Stage 1: Prior to delivery	Stage 2: Delivery, unloading and handling on site	Stage 3: Fixing	Stage 4: Post-fixing and prior to handover
Design incompatible with standard components encourages waste.	Materials broken on delivery	Waste resulting from cutting materials to size	Vandalism
Materials wrongly specified when ordering – goods rejected by clerk of works or materials of higher quality than required.	Materials damaged in unloading and stacking	Negligent use of materials	Inadequate protection of completed work
Incorrect quantities ordered – errors in preparing quantities or ordering direct from an inaccurate B.Q.	Damage due to multiple handling around the site.	Surplus materials on completion of operation.	Damage caused by other trades.
	Damage or loss of materials due to poor storage conditions	Substitution of a material for a more expensive item.	
	Theft of materials	Defective work to be re-done or corrected.	
	Short delivery of materials or excess delivered, resulting in an unacceptable level of surplus.	Work not built in strict accordance with drawings	

5. Allowances for the method of measurement of the bill item
Clause A.3.2 of SMM 6 states, *inter alia*, that work shall be measured 'net as fixed in position' which means that in certain circumstances a larger quantity of materials, irrespective of waste, must be purchased in order to produce the measured quantity in the bill, the corresponding cost of course being incorporated into the bill rate. It is therefore essential that the estimator understands how the quantity surveyor measures building work generally and can recognise particularly those bill items which are governed by this rule.

Typical examples of where allowances must be made for the 'net' measurement of quantities are as follows:

- *Allowance for laps in the length or area* – applicable to DPC, fabric reinforcement, lead flashings, roof tiles and slates, etc.
- *Compaction of materials when laying* – hardcore, floor screeds, etc.
- *Shrinkage occurring in mixing* – concrete, plaster, mortar, etc.
- *Bulking in excavation* – the quantity of material to be disposed of after excavation will be greater than that stated in the bill.

4.3 Mechanical Plant Cost

A great variety of mechanical plant is now available for use in building work to speed up operations and to tackle otherwise impossible jobs. However, the cost of most items of mechanical plant and equipment is very high and, before authorising their use on any building contract, the contractor must be satisfied that the utilisation of any piece of plant is worth his while and will in the long run produce savings in time and/or money.

Each area of work where there exists a possibility of being able to use mechanical plant must be carefully considered by the estimator, probably in conjunction with the plant or contracts manager, at the tender stage and decisions made with regard to the nature of the equipment to be used and the length of time it will be needed on site. The preparation of a preliminary bar chart or critical path programme would assist in this decision making.

4.3.1 The provision of mechanical plant (ownership v hiring)

Traditionally, contractors have generally owned their own plant and equipment either directly or through a subsidiary company established to deal specifically with this aspect of the company's activities, with items of plant being bought outright either new or second-hand or by hire-purchase agreement or other credit terms. However, with the ever-increasing cost of mechanical plant and the traditionally low liquidity ratio in the industry, the trend in recent years has been towards a reduction in the stock of expensive plant held directly by contractors and an increase in the number of specialist plant hire companies set up to meet the still present demand for mechanical plant and whose sole business is to cater for the needs of the industry in this respect.

The relative advantages and disadvantages of owning and hiring plant may be summarised as follows.

Advantages of ownership

1 *Convenience*. Despite the proliferation of the specialist hire firms, most contractors still carry certain items of plant which are considered to be versatile and capable of being used to the full with the minimum amount of time being spent idle in the contractors' yard. Perhaps the main advantage of the ownership of plant is the convenience of having it on hand when it is needed at short notice without having to place an order with a plant hire firm and maybe having to wait a day or two for delivery to site.

2 *Plant can be purchased for particular purposes*. Where contractors choose to own particular items of plant, a decision on which type or model to buy can be based on the company's specific needs and, accordingly, the most suitable and efficient item of plant purchased without the need for compromise.

3 *Flexibility of pricing*. Where the plant is owned, the contractor is able to enjoy a certain amount of flexibility in his pricing policy and on occasions may wish to include any mechanical plant within his tender at net cost in order to maintain a competitive price. However, this flexibility is lost where plant is hired, since any rates would always include an element for the hire company's overheads and profit.

4 *Flexibility of operation*. When mechanical plant is not in use by the contractor there may be scope for hiring it out to other companies on a commercial basis, perhaps even in competition with the hire companies themselves, thereby creating an opportunity for a further source of revenue for the firm.

5 *Tax allowances, grants and subsidies*. From an accounting point of view, there may be sufficient incentives to encourage the contractor to embark on capital expenditure. For example, the Government, through its regional and national incentive schemes, provides regional development grants for capital expenditure on certain items of plant and machinery where firms have either moved into a designated development area or who can expand or modernise an existing operation within one of these areas. Through the system of national taxation allowances, capital expenditure on plant and machinery anywhere in the UK can qualify for a first-year tax allowance of 100 per cent. Other grants and subsidies may also be available depending on current Government policy in this field.

Disadvantages of ownership

1 *Difficulties in keeping up to date with equipment.* An important objective of any manufacturer of mechanical plant is the constant modernisation and improvement of his product resulting from advances in design and technology in order to produce a more efficient machine in an attempt to maintain or improve sales. Therefore, a contractor can, having spent a large sum of money on a piece of plant, find that it has become out of date very quickly, perhaps long before the end of its useful life, presenting the contractor with the dilemma of continuing to use a now relatively inefficient piece of equipment or of replacing it at great expense. The plant hire company, dealing solely in the purchase of plant, would be better placed to keep up to date with the latest models, probably having more resources available for this purpose than the contractor and perhaps being in a position to negotiate substantial discounts with manufacturers in return for regular large orders. By keeping up to date with its stock of plant and equipment, the plant hire company can maintain its appeal within the industry.

2 *Ties up capital.* Despite the availability of grants and tax allowances, the contractor must still be able to find substantial sums of money for the purchase of mechanical plant and, by committing a large part of his resources to this activity, may find himself being severely restricted in other areas through a shortage of capital.

All building work requires heavy financing, especially in the initial stages, and the contractor must have sufficient funds on hand to meet immediate demands in order to avoid costly borrowing from the banks.

3 *The costs of maintaining plant.* Where a reasonable stock of plant is owned by a contractor, it must be kept in good running order by a plant maintenance depot which would be responsible for the day-to-day servicing and repair of items of machinery. The considerable cost of running such a department with its skilled labour force, accommodation, tools, stock of spares, etc. must be carefully weighed against the advantages of owning plant, as the whole system could be dispensed with if all plant and machinery were to be hired.

The cost of plant when not in use. In situations where mechanical plant is hired, the contractor will keep the machine only for as long as it is actually needed on site, despatching it back to the hire company promptly as quickly as possible whence the hire charges will cease. However, where plant is owned, the now redundant piece of equipment must be brought back to the contractor's plant depot where it may stand for some weeks before being needed again, and although the estimator will have allowed for a certain amount of idle time within his all-in rate build-up, any standing time in excess of this allowance will result in the contractor failing to recover the full cost of operating the machine.

Since the estimator's allowance for lost time is likely to be minimal in order to keep his prices competitive, and since it is unlikely that the contractor will be able to find alternative work for the machine's operator (where applicable), any mechanical plant which is not in use and lying idle in the contractor's yard will prove an expensive luxury.

5 *The cost of transporting plant to and from site.* With the exception of wheeled excavators and lorry-mounted cranes, all other items of mechanical plant need hauling to and from the site on low loaders or other suitable vehicles. The cost of providing this service is likely to be considerable, especially where an item of plant is needed for only a short time when the cost of transportation can often exceed the cost of the plant itself, perhaps even rendering the use of mechanical plant uneconomic in that instance. When mechanical plant is hired, however, the cost of transporting the item to site is normally included in the hire rate.

4.3.2 The main features of hiring mechanical plant

When mechanical plant is hired, the contractor is required to enter into a hire agreement with the hire company for the item of plant involved which then binds him to the terms and conditions which are normally set out on the back of the form of agreement and, whilst the details of such hire agreements will differ from one firm to another, the following conditions would be typical of most agreements.

Normal conditions of hire

1 *Hire period.* The hire commences when the item of plant leaves the owner's depot and ends when it is received back at the depot after use.
2 *Working period.* This is the hire period less the amount of time spent in travelling both ways on the public highway. This is particularly important where plant travelling under its own power is involved since the hire rates whilst travelling and working are often different.
3 *Time sheets.* For items such as cranes and excavators, where an operator is provided, the contractor is required to sign daily or weekly time sheets signifying his acceptance of the number of hours spent on the job. These time sheets will form the basis of the subsequent charges and it is the contractor's responsibility to find sufficient work for the machine on site to keep it fully

operational for the time it is on hire, since the hourly or daily hire charges will not be suspended for periods when there is no work available.

4 *Site conditions*. The contractor is responsible for ensuring safe working conditions for the machine whilst on site, including the preparation of the ground by providing hardstanding areas and temporary access roads where necessary to take the weight of cranes and other heavy items of plant. Further, the contractor is liable for any costs arising out of loss or damage to the mechanical plant through soft ground conditions, etc. and also for puncture and tyre damage (with the exception of fair wear and tear or defective tyres). In order to protect himself against his liabilities, therefore, the contractor must arrange for adequate insurance cover although, in practice, his normal blanket policy for site working would probably include hired plant.

5 *Operation of plant*. Where an operator is provided with the machine, he is deemed to be a servant of the contractor and must act on instructions given on the site. This makes the contractor responsible for the actions of the operator and his machine, including any damage caused to underground services, adjacent property, etc. (unless the operator deliberately disregards specific instructions). The contractor must therefore take great care to locate the position of all existing drains, gas and water mains and electricity cables and take all necessary precautions.

6 *Breakdowns*. Where an item of plant breaks down or is in need of adjustments or servicing, the owner undertakes to carry out the necessary repairs as quickly as possible or replace the item with another one of similar type and condition.

On occasions where the owner is unable to provide a substitute, the hire is normally terminated, but it should be noted that he will not usually accept any consequential loss by the contractor resulting from the breakdown, stoppage or delay.

7 *Notice of termination*. The owner usually requires notice from the contractor of the termination of hire – four days being a typical period demanded.

Irrespective of the relative advantages and disadvantages of owning and hiring mechanical plant, it is likely that over the years a contractor would have acquired the more commonly-used items of plant whilst relying on hired equipment for specialist tasks. Such common items of mechanical plant would include:

Mixing plant
Dumpers
Compressors and tools
Hoisting and lifting equipment
Roller and punners
Portable saw bench
Disc cutters
Lorries and vans
Wheeled/tracked excavators
Pumps and hoses
Concrete vibrators
Generators

4.3.3 The incorporation of mechanical plant costs within the tender

Having decided upon his needs for the mechanical plant to be used on the contract and the length of time each item would be required on site, the estimator must then calculate the total cost for incorporation within the tender figure. With the introduction of SMM 6 in 1979, the estimator is now given the opportunity of including such costs in one, two or all of the three sections specifically provided within the bill of quantities for this purpose.

The plant costs thus calculated may be included in the bill:

1 Within the bill rates themselves

Section A.4 of the general rules states under paragraph 2.d that, unless stated otherwise, all items within the bill shall be deemed to include, *inter alia*, 'Plant and all costs in connection therewith'. In this case the assumption is that plant costs are related directly to quantity, since they will be calculated on the basis of 'per unit of measurement' for the particular operation described. This method of pricing mechanical plant is particularly suited to those trades where a piece of equipment can be readily identified with a specific operation thereby enabling the full, true cost to be allocated to that part of the work. Examples showing plant costs included within the bill rates are given later under 'Concrete work', 'Excavation', 'Brickwork and Blockwork', etc.

The following extract from SMM 6 illustrates this method.

A.4 Descriptions

1. The order of stating dimensions in descriptions shall be consistent and generally in the sequence of length, width and height. Where that sequence is not appropriate or where ambiguity could arise, the dimensions shall be specifically identified.
2. Unless otherwise specifically stated in the bill or herein the following shall be deemed to be included with all items:
   a. Labour and all costs in connection therewith.
   b. Materials, goods and all costs in connection therewith.
   c. Fitting and fixing materials and goods in position.
   d. Plant and all costs in connection therewith.
   e. Waste of materials.
   f. Square cutting.
   g. Establishment charges, overhead charges and profit.
3. Junctions between straight and curved work shall in all cases be deemed to be included with the work in which they occur.
4. Notwithstanding the provisions in this document for labours to be given as linear items, such labours may be given in the description of any linear items of work on which they occur.
5. Notwithstanding the provisions in this document for labours to be enumerated, such labours may be given in the description of any enumerated item of work on which they occur.

2. As a lump sum in the Preliminaries section

Under the sub-heading of 'General facilities and obligations', and for convenience of pricing, paragraph B13 1.a of the standard Method headed 'Plant, tools and vehicles' provides an opportunity for the estimator to include his plant costs here if he wishes. He may choose to include the plant costs for the whole contract here, or only those items whose cost cannot accurately be allocated to any particular part of the work, but are used more or less continuously throughout the contract by several trades. Examples of such items would be: scaffolding, dumpers, hoists and lifting equipment, etc.

The following extract from SMM 6 illustrates this method.

GENERAL FACILITIES AND OBLIGATIONS

B.13 Pricing

1. For convenience in pricing, items for the following shall be given. Maintaining temporary works, adapting, clearing away and making good shall be deemed to be included with the items. Notices and fees to local authorities and public undertakings related to the following items shall be deemed to be included with the items.
   - **a. Plant, tools and vehicles.**
   - b. Scaffolding.
   - c. Site administration and security.
   - d. Transport for workpeople.
   - e. Protecting the works from inclement weather.
   - f. Water for the works. Particulars shall be given if water will be supplied by the Employer.
   - g. Lighting and power for the works. Particulars shall be given if current will be supplied by the Employer.
   - h. Temporary roads, hardstandings, crossings and similar items.
   - j. Temporary accommodation for the use of the Contractor.
   - k. Temporary telephones for the use of the Contractor.
   - l. Traffic regulations.
   - m. Safety, health and welfare of workpeople.
   - n. Disbursements arising from the employment of workpeople.
   - p. Maintenance of public and private roads.
   - q. Removing rubbish, protective casings and coverings and cleaning the works on completion.
   - r. Drying the works.
   - s. Temporary fencing, hoardings, screens, fans, planked footways, guardrails, gantries and similar items.
   - t. Control of noise, pollution and all other statutory obligations.

3. As a lump sum within each work section

With the exception of woodwork, metalwork, glazing, painting and fencing, the Standard Method provides, within each work section, items against which the estimator may include the anticipated plant costs applicable to that particular trade where he is able to identify with any degree of accuracy, the costs involved.

Examples of these measured items can be found in the Standard Method under clauses C2, D4, E3, F2, G2, etc. where **two** items are taken into account:

(a) The cost of transporting the items of plant to and from the site.
(b) The cost (hire or otherwise) of the items of plant whilst remaining on site and including all attendance and maintenance costs in connection therewith.

The introduction of these 'plant' items within the work sections imply that the cost of mechanical plant is not merely 'quantity' related as method (a) above suggests, but 'time'-related which is, in practical terms, much nearer to reality!

Traditionally, contractors have been extremely wary of divulging too much information with regard to the actual make-up of individual elements of the tender figure, and it may be that the estimator will feel much safer 'hiding' behind 'all-embracing' large lump sums in the Preliminaries which can inevitably cover a 'multitude of sins'! However, the estimator is given the choice of methods of pricing and he has the freedom to select whichever method(s) he feels is appropriate to the various parts of the work.

The following extracts from SMM 6 and the practice manual illustrate the thinking behind this method.

PLANT ITEMS

The Standard Method of Measurement requires items to be given in the bill for provision and maintenance of plant on site. The reason for giving plant separately from the ordinary items for permanent work is in recognition of the fact that plant costs often depend upon factors other than the quantities and nature of the various items of permanent work in the Section. For example, in erecting precast concrete, the plant cost would include the cost of a crane for erecting precast concrete units. The cost of the crane would depend upon what size of crane was provided and how long it was likely to be required on site. The choice of the crane would depend upon the heaviest duty required, usually the heaviest unit at greatest radius and this would govern the daily charge for keeping the crane on site and the charges for haulage to and from site. The length of time would be governed by other factors such as the rate at which the work becomes available and the number and type of precast units as a whole. The total plant cost consequently has a fixed element (the cost of providing the plant) and a time related element (the cost of maintaining the plant on the site for the duration required). These elements of cost are normally considered separately when estimating and it is helpful if they can also be considered separately when valuing variations.

The plant items enable these cost elements to be priced separately instead of being spread over other groups of items which should contribute to more realistic pricing generally. It is not the intention that tenderers be forced to vary their method of pricing. If the items are left unpriced the plant cost will be assumed to have been included elsewhere.

Section F

CONCRETE WORK

F.2 Plant

1. An item shall be given for bringing to site and removing from site all plant required for this section of the work.
2. An item shall be given for maintaining on site all plant required for this section of the work.

Example illustrating the pricing of the mechanical plant element required for the concrete work section using the above method:

				£	p
A	Allow for bringing to site all plant required for this section of the work and removing on completion	Item			
B	Allow for maintaining on site all plant required for this section of the work	Item			

Assumptions:
Plant requirements: (from method statement and programme)

2no. 14/10 concrete mixers for 8 weeks
3no. vibrator pokers for 8 weeks
2no. power floats for 2 weeks

Item A
Assume two return journeys are needed for transporting the mixers and vibrators.
Assume one return journey for transporting the power floats.

(a) Mixers and vibrators:

loading plant at yard –	1 hour for 4 labourers at £4.00 per hour	=	£16.00
	1 hour for lorry and driver at £12.00 per hour	=	12.00
travelling to site –	1 hour for lorry and driver at £12.00 per hour	=	12.00
unloading at site –	$\frac{1}{2}$ hour for 4 labourers at £4.00 per hour	=	8.00
	$\frac{1}{2}$ hour for lorry and driver at £12.00 per hour	=	6.00
returning to yard –	$\frac{3}{4}$ hour for lorry and driver at £12.00 per hour	=	9.00
			£63.00
Allow for bringing plant back from site on completion of work, say			63.00
		Total	£126.00

× 2 for two journeys = £252.00 . . . 1

(b) Power floats:

loading plant at yard –	¼ hour for 2 labourers at £4.00 per hour	=	2.00
	¼ hour for lorry and driver at £12.00 per hour	=	3.00
travelling to site –	¾ hour for lorry and driver at £12.00 per hour	=	9.00
unloading at site –	¼ hour for 2 labourers at £4.00 per hour	=	2.00
	¼ hour for lorry and driver at £12.00 per hour	=	3.00
returning to yard –	¾ hour for lorry and driver at £12.00 per hour	=	9.00
			£28.00
Allow for bringing plant back from site on completion of work, say			28.00
	Total		£56.00 . .2

Total cost for item A (1 + 2) = £308.00

Item B

Hire costs:

2no. 14/10 mixers say at £60.00 per week for 8 weeks	=	960.00
3no. vibrator pokers at say £20.00 per week for 8 weeks	=	480.00
2no. power floats at say £25.00 per week for 2 weeks	=	100.00
Allow for fuel, oil, grease etc. say £100.00	=	100.00
Allow maintenance, say 1 hour per day for 8 weeks at £4.20 per hour (including NWRA allowance)	=	168.00
Total		£1,808.00

Total cost for item B = £1,808.00

4.3.4 The build–up of plant costs

Whichever of the above methods of including plant costs in the tender are used, the estimator will need to calculate the total hourly cost of using each item of plant 'in the field', the procedure varying depending upon whether the plant is owned or hired as illustrated in the following examples.

1. Building up an hourly rate where plant is hired

Whether the plant hirer charges by the hour, day or week for mechanical plant on hire, the estimator must convert the charges to a cost per productive hour on site making due allowances for:

(a) Standing time.
(b) Maintenance (i.e. oiling, greasing and cleaning down at the end of the day).
(c) Fuel and oil.
(d) The cost of an operator (where applicable).

Example: To calculate the effectively hourly cost of operating a $5/3\frac{1}{2}$ mixer on site when the weekly hire charge is £35.00.

Assumptions:

1 The normal-five day 39-hour week is worked.
2 Allow 1 hour per day standing time due to stoppages, bad weather, etc.
3 Allow $\frac{1}{2}$ hour per day maintenance time.

Total number of productive hours per week $= 39 - 5 \times (1 + \frac{1}{2})$
$= 39 - 7\frac{1}{2}$
$= 31\frac{1}{2}$

Therefore effective hourly hire rate of mixer $= \dfrac{£35.00}{31\frac{1}{2}}$ = £1.11

Fuel and oil – allow 1 litre of fuel at 35p per litre = 0.35
allow, say, 10p for oil, grease, etc. = 0.10

Operator – assuming an all-in rate of £4.00 per hour plus 10p for operating mechanical plant (NWRA) = £4.10

Cost of maintenance – $\frac{1}{2}$ hour per day = $2\frac{1}{2}$ hours per week at £4.10 per hour = £10.25 per week ∴ cost per productive hour $= \dfrac{£10.25}{31\frac{1}{2}}$ = 0.33

Total cost of operating per hour = £5.99

Notes:

(a) NWR 7 provides for the operators of mechanical plant, where applicable, to be allowed time within the working day to oil, grease, clean down and generally prepare the piece of equipment ready for use the next day.
(b) NWR 3 provides for extra payments above the normal labourers' rate to be made to operatives who are given the responsibility of driving or operating mechanical plant.
(c) The calculation assumes that alternative work can be found for the operator when the machine is lying idle. Where the estimator feels that this will not be possible, the extra cost of the operative's non-productive hours must be included.

The pricing principle illustrated can be extended to most items of mechanical plant though the allowances for standing time, operating and fuel costs would of course differ in each case, the estimator calling on his experience and advice from others to determine actual figures.

2. Building up an hourly rate where plant is owned

Where mechanical plant is owned, the build-up of an hourly all-in rate must take into account the full costs of ownership including the purchase price and depreciation in addition to the operating costs as before. The factors to be considered are as follows:

(a) Purchase price of the machine.
(b) Allowance for depreciation.
(c) Standing time.
(d) Cost of repairs and renewals.
(e) Fuel and oil.
(f) Insurance and licensing costs (where applicable).
(g) The cost of an operator or attendant (where applicable).

Example: To calculate the effective hourly cost of owning and operating a 1 m³ dumper on site when the purchase price is £3,000.00.

Assumptions:
1. The dumper has a useful life of 4 years before it is scrapped.
2 Allowing for standing time, holidays and lying idle between contracts, the number of actual working hours on site is reduced to, say, 1,5000 per year.

Initial cost of machine	£3,000.00
Less scrap value, say,	150.00
Therefore depreciation over 4 years	£2,850.00
Add interest on capital, say 10%	285.00
Therefore total cost over 4 years	£3,135.00

Total number of hours worked over 4 years = 1,500 × 4 years
= 6,000 hours

Therefore basic hourly cost	=	$\frac{£3,135.00}{6,000}$
	=	0.52
Repairs and renewals – allow 15%	=	0.08
Fuel and oil – allow 1½ litres per hour at 35p per litre	=	0.53
Insurance – allow £75 per annum:		
Hourly rate = $\frac{£75}{1,500}$	=	0.05
Licence cost – NIL	=	—
Driver – Assuming an all-in rate of £4.00 per hour plus 20p (NWRA)	=	4.20
Allow ½ hour per day maintenance, labour cost = $\frac{£3.48 \times \frac{1}{2}\text{ hour}}{8\text{ hours}}$	=	0.22
Total cost of owning and operating dumper per hour		£5.64

The cost of transporting items of mechanical plant to and from the site is dealt with under Preliminaries and assumes that a lump sum is included in this respect in this section of the bill or under the respective 'plant' items within each work section where specific plant costs can be accurately allocated to individual trades. Alternatively, however, the transport costs for each item of plant could be calculated and subsequently incorporated into the hourly all-in rate. The choice of method lies with the estimator.

4.4 Attendance

Work of a specialist nature which is to be carried out by a nominated sub-contractor (i.e. a firm selected by the architect) is included in the bill of quantities as a prime cost sum and is thereby incorporated into the tender figure together with a separate percentage addition for the main contractor's profit should the estimator wish to price this item.

However, in addition to the work carried out directly by the nominated sub-contractor, further additional work is often needed in the form of facilities to be provided by the main contractor on behalf of the sub-contractor whose scope of operations frequently does not extend to the provision of certain 'on-site' requirements.

These facilities which are provided are known as ATTENDANCES, the nature and extent of which will vary depending on the type and scope of the specialist work involved on the contract.

As the bill of quantities is being prepared, the quantity surveyor will examine each type of specialist work for which a prime cost (PC) sum is to be included in the bill and identify those items of attendance which he knows would be required by the nominated sub-contractor. Such items must then be specified and measured as 'items' immediately following the corresponding PC sum and its profit, these items of attendance falling into two distinct categories in accordance with clause B.9 of the Standard Method:

1 B.9.2 General Attendance.
2 B.9.3 Other (or special) Attendance.

4.4.1 General Attendance

Refers to the provision by the main contractor to the nominated sub-contractor of his existing services and facilities which he has already installed for his own use and convenience, e.g. temporary roads, standing scaffolding, temporary lighting, water supplies, mess rooms, etc. Such facilities are known as general attendances and are to be provided by the contractor free of charge. Normally the estimator will have priced for such items in the Preliminaries and usually no extra cost will be incurred, although the item of general attendance must still be measured in order to give the contractor the opportunity of including for, say, additional wear and tear on roads, scaffolding, etc. However, it would be almost impossible to identify such extra over costs which at the most would be of an extremely minor nature and in practice the general attendance item is very rarely priced.

4.4.2 Other (special) Attendance

Refers to the provision of specific services by the contractor to the nominated sub-contractor which are identifiable and which can be seen to impose an obvious additional cost upon the contractor.

Examples of such services are given in clause B.9.3 of the Standard Method under items (a) – (g). Where one or a number oi these sort of items are deemed necessary by the quantity surveyor they will be measured in the bill and the estimator will almost certainly wish to insert a price, the cost depending upon the nature and extent of the resources to be provided.

The method of presentation and pricing of attendance items in the bill of quantities would be as follows:

Bill No. 1 – PROVISIONAL AND PRIME COST SUMS

				£	p
	PRIME COSTS SUMS				
	Nominated Sub-contractors Provide the following sums for work to be executed complete:				
A	Steel portal framed structure			*7.500*	*00*
B	Profit		*2½%*	*187*	*50*
C	General Attendance	item	–	–	–
D	Other Attendance				
	Providing hardstanding area 4 m wide to North and South elevations to accommodate 6 tonne mobile crane	item		*1,942*	*00*
	Covered storage accommodation, min. size 5 m × 3 m on plan	item		*441*	*00*
	Protecting when completed	item	–	–	–
E	Heating installation complete			*15.000*	*00*
F	Profit		*2½%*	*375*	*00*
G	General Attendance	item	–	–	–

H	Other Attendance Unloading, distributing and storing materials and equipment as follows: 1 oil storage tank 2,440 × 1,520 × 1,220 1 oil fired boiler 60,000 BthU rating 2 accelerator pumps 20 double panel radiators 1,600 × 600 200 m (approx.) copper pipe and fittings 1 water storage tank 1,520 × 1,140 × 910 1 hot water storage unit 1,000 × 2,000 pipe insulation, lagging, sundries	item			*58*	*00*
	Covered storage accommodation, min. size 10 m × 3 m on plan	item			*660*	*00*
	Electricity for power tools	item			*220*	*00*
	Protecting when completed	item		–	–	–
	Bill no. 1 carried to summary				*26,383*	*50*

Method of pricing

The above items may be priced using the following hypothetical assumptions:

1. Profit

Assume: the contractor's policy is to include for 2½ per cent profit on all nominated sub-contractor's work.

2. General Attendance

Assume: no additional costs would be involved and therefore item not priced.

3. Other (special) Attendance

(a) *Provision of hardstanding*

From the drawing, assume the length of the North elevation is 50 m. Therefore, total area of hardstanding required = 2 × 50 m × 4 m = 400 m^2.

Assume hardcore bed is to be 200 mm thick, therefore total volume = 400 m^2 × 0.2 = 80 m^3 at, say, 1½ tonne per m^3 = 120 tonnes. Allow for 25 per cent compaction and waste = 30 tonnes. Therefore total amount required = 150 tonnes at, say, £9.00 per tonne = £1,350.00

(i) *Labour and plant*: Assume: 1 front bucket loader and gang of 3 labourers take 2 days to lay, spread and consolidate the hardcore bed.

Assume all-in rate for loader and driver = £12.00 per hour
Assume all-in rate for labourer = £4.00 per hour

Therefore hourly cost of operation	= £12.00 + 3 × £4.00 = £24.00
Total cost of laying	= 16 hours × £24.00 per hour
	= £384.00

(ii) *Removing on completion*: Assume: 1 front bucket loader, 1 lorry and driver, and 1 labourer required for 1 day. Assume all-in rate for lorry and driver = £10.00 per hour.

Therefore hourly cost of operation
= £12.00 + £4.00 + £10.00 = £26.00. Total cost of removing hardstanding area
= 8 hours × £26.00 per hour
= £208.00

Total cost of special attendance item
= £1,350.00 + 384.00 + 208.00
= £1942.00

(b) *Provision of storage accommodation*

Assume: hire rate of 5 m × 3 m site hut is £30.00 per week.
Assume: hut is required for 8 weeks, therefore hire cost = £30.00 × 8 = £240.00. Loading and transporting hut to site:

2 hours' hire of lorry and driver at £10.00 per hour	= £20.00
2 hours for 2 labourers at £4.00 per hour	= 16.00
	£36.00

Allow for loading and transporting hut back from site after use = £36.00

Total cost of transport and loading = £72.00

Erection and striking of hut on site:
Assume: gang of 3 joiners and 2 labourers for 6 hours in total, hourly cost
= 3 × £4.50 + 2 × £4.00 = £21.50 × 6 hours = £129.00

Total cost of providing accommodation
= £240.00 + 72.00 + 129.00
= £441.00

(c) *Protection of work when completed*
Assume: not priced by main contractor.

(d) *Unloading and distribution of equipment and materials*
Assume: fork-lift truck required for 2 hours at say,
£11.00 per hour = £22.00

Assume: gang of 3 labourers required for 3 hours,
= 3 × £4.00 × 3 = £36.00

Total = £58.00

(e) *Provision of 10 m × 3 m site hut*
Principle involved in calculation as above, hire rate and labour cost would be higher due to larger hut required. Say, half as much again – £660.00

(f) *Provision of electricity for power tools*
Provision of cables, leads, plugs, etc., say, £100.00

Assume: 200 units of electricity per week used at 10p per unit for 6 weeks
= 200 × 10p × 6 weeks = £120.00
Total cost = £100.00 + 120.00 = £220.00

Chapter 5

Contract clauses and their effect on cost

One of the most important documents involved in the building process is the form of contract. It provides the basis of the agreement between both parties who are legally bound by the terms and conditions embodied within its clauses.

It is unlikely that the form of contract will be despatched with the other tender documents to contractors at the tender stage; however, a 'schedule of clauses headings' which are to apply is incorporated into the preliminaries bill together with details from the appendix to the conditions.

It is vital therefore that the estimator has a thorough working knowledge of the contract and the implications upon his tender of every clause.

The form of contract provides for certain clauses to be deleted and for particular conditions to be written into the appendix. Each job the estimator prices, therefore, will contain different provisions, and misinterpretation at this stage of the employer's requirements could be disastrous, resulting in the contractor being unsuccessful with his tender or worse, obtaining the contract but subsequently being doomed to lose money on the job. Although it will be appreciated that there are many different types of building contract, this chapter is related to the JCT 'Standard Form' of contract for use with quantities, 1980 edition. However, the GC/WKS/1 contract for use on Central Government contracts and the ICE conditions for use in the civil engineering field are similar in principle.

The form of contract is constantly being revised and, again, it is essential that the estimator keeps up-to-date with all the latest amendments.

This chapter deals with those clauses of the JCT form which are likely to have a direct effect on cost, and which must be carefully considered by the estimator when compiling his tender.

The clauses dealt with are:

1	Clause 10	Person in Charge (foreman or site agent).
2	Clause 15	Value Added Tax.
3	Clause 17	Practical Completion and Defects Liability.
4	Clause 18	Partial Possession by Employer.
5	Clause 19A	Fair Wages.
6	Clauses 20, 21 and 22	Insurances.
7	Clause 23	Possession and Completion.
8	Clause 24	Damages for Non-Completion.
9	Clause 25.4.10.1. 25.4.10.2	Extension of Time – Availability of Labour, Goods and Materials
10	Clause 30	Certificates and Payments
11	Clauses 38, 39, 40	Fluctuations

5.1 Clause 10 – Person in Charge (foreman or site agent)

Clause 10 calls for the contractor to keep on the site at all times a competent person/foreman in charge, and although the contract does not specifically state this, the bills of quantities often say that the foreman shall not be changed without the consent of the architect.

Quite often a contractor will employ a bricklayer foreman on the site during the early stages of a contract and then change to a joiner foreman or one experienced in the finishing trades to complete the project. (If the above provision is stated in the bill, then the architect's consent must be obtained before the foreman can be changed.)

Although this move can be of benefit to the contractor, it must be remembered that quite often oral instructions given to the foreman are not always confirmed straight away by the architect, the foreman carrying the information around in his head until he remembers to contact the contracts manager or quantity surveyor. Thus a change in the foreman mid-way through a contract could result in information regarding variations not being passed on.

The cost of the foreman, together with the costs of all other site and supervisory staff such as trades or section foremen, gangers, storeman, site clerks, etc., will normally be priced under the heading of 'Person in charge' in the preliminaries bill.

The cost of site staff will obviously depend on the size and complexity of the job and the contract period.

On a large involved project there may be a site-based contracts manager together with one or more assistant foremen and perhaps a number of trades foremen and gangers. A job of this size would also require a site clerk to check the delivery of materials, prepare labour and plant returns, and perhaps assist in labour calculations. A storeman would probably be needed to take charge of and keep a record of all sundry materials used.

A site-based quantity surveyor and bonus surveyor may also be employed, but their costs would probably be included in the Head Office overheads. Consider a contract period of 18 months (89 weeks) for a large housing project. A typical build-up for the supervision item, Clause 10, could be as follows:

Assume the foreman's salary is £8,000 per annum and in addition he has a company car and is allowed all expenses and fuel, etc. in connection therewith. (*Note*: the total cost for the full 89 weeks must be allowed since he will still be paid during holiday periods.)

General foreman

89 weeks at £154 per week	£13,706.00	£21,271.00
Provision of company car: 89 weeks at say £60 per week plus fuel, tax, servicing, etc., 89 weeks at, say, £25 per week	£7,565.00	

Trades foreman

Assume the trades foremen are working in a supervisory capacity for 50 per cent of their time, the remaining time being productive (i.e. laying bricks, fixing roof joists, etc.) and therefore being included in the bill rates

2 Bricklayer foremen for half the contract: 2 × 45 weeks at £120 per week × 50 per cent	£5,400	£12,600.00
2 Joiner foremen for two-thirds of the contract: 2 × 60 weeks at £120 per week × 50 per cent	£7,200	

Gangers

Assume the gangers are working in a supervisory capacity for only 25 per cent of their time

2 gangers for 20 weeks 2 × 20 weeks at £100 per week × 25 per cent.	£1000.00

Site clerk

The site clerk would probably not be required for the full contract period – allow, say, 80 weeks

80 weeks at £90 per week	£7,200.00

Storeman

1 storeman 89 weeks at £90 per week £8,010.00

TOTAL COST £50,081.00

(Note that the costs of the site-based quantity surveyor and bonus surveyor are not included here, but are allowed for in the overhead percentage.)

The salaries and wages quoted in this example should include the cost to the contractor of employing these men, i.e. the 'all-in' cost of employment. Alternatively, these 'labour-on' costs could be calculated separately and included as a lump sum under that particular heading in the preliminaries section of the bill.

5.2 Clause 15 – Value Added Tax

With regard to the implication upon a tender of Value Added Tax, Clause 15.2 states that the contract sum shall be exclusive of any VAT which might be applicable on the contract. It can be said, therefore. that although VAT may subsequently be paid by the employer on part or on the whole of the contract, the effect of the VAT clause upon the tender figure is zero.

However, Practice Note 17, issued by the Joint Contracts Tribunal in support of the JCT Standard Form, recommends that since the tender sum will be VAT exclusive, employers may not at this stage be aware of their likely liability for the tax payable. It is, therefore, suggested that as a benefit to the employer, the contractor should be asked to submit a provisional assessment of the amount of tax which may become due (such provisional assessment would not be binding on either party).

If this recommendation is followed, the task of assessing the approximate amount of VAT will probably fall within the estimator's province, since he will have priced the bill and will be familiar with the nature of the total cost. It may be as well, therefore, to list briefly in note form the effect of VAT upon the building industry.

1 VAT was introduced on 1st April 1973 under the provisions of the Finance Act 1972 and is collected by HM Customs and Excise.
2 It is a tax levied on the supply of goods and services supplied by contractors in the normal course of their work.
3 The rate of tax is either 'zero' or 15 per cent (the existing 'positive' rate). This rate being subject to alterations at any time due to Government legislation.
4 Building work can be either 'zero' or 'positive' rated or a mixture of the two depending on the nature of the work. All new construction is zero rated, whilst all other building work including extensions, repairs and alteration work is positive rated.

5 The amount of tax, where applicable, is assessed by the employer as the work proceeds and is subsequently invoiced separately by the contractor, but at the same time as each interim certificate. The amount paid is then passed on by the contractor to the Customs and Excise Department. (*Note*: It is the contractor who is liable for the tax and it is his responsibility to recover the amount from the employer.)
6 The contractor, when purchasing materials for use in building work, must pay VAT at the positive rate to his suppliers, even though the materials could subsequently be incorporated in zero-rated work. In such cases, he cannot recover the amount of VAT he has paid from the employer, but must claim the sum back from the Customs and Excise, who will refund the amount previously paid. (The contractor's own accounts department will deal with all the invoices, records, etc. relating to the paying and refunding of the tax to and from the Customs and Excise.)

Immediately following the Appendix in the Standard Form there are 'Supplemental Provisions',otherwise known as the VAT Agreement which legally binds the employer to pay to the contractor the correctly assessed amount of tax. Remember, it is the contractor who is responsible for paying the tax, and this agreement allows him to recover this sum of money from the employer.

To summarise then, the work which the estimator may be called upon to do at this stage is to estimate the value of that part of the work which is 'positive' rated and calculates 15 per cent of that sum. This amount will be thc provisional total of VAT to be charged.

5.3 Clause 17 – Practical Completion and Defects Liability

Under this clause the contractor is required to make good any defects arising out of bad workmanship or the use of materials not in accordance with the contract (see Clause 8), which become apparent during the defects liability period. (This being the period of time, usually lasting 6 months, which commences on the date of the issue of the 'Certificate of Practical Completion' by the architect.) Within 14 days of the expiration of the defects liability period, the architect should issue to the contractor a schedule of defects outlining the work to be remedied.

Obviously, the contractor will be involved in costs in making good the defects, costs which cannot be recovered under the contract unless the estimator has included a provision for this item within the original tender figure. How much to allow? If anything at all! Indeed, some estimators leave this item unpriced, hoping that the supervision on site will be good enough to keep the standard of workmanship high, thereby keeping the amount of defects down to a minimum so as to be insignificant and not worth pricing. If the contract is properly organised and all goes well, sufficient profit may be made over and above that originally budgeted for so that any such costs involved could be set against that.

However, the contractor may not wish to pay for the making good of defects out of his hard-earned profit; indeed, when work is scarce, the anticipated profit margin may be very small or even non-existent. In such cases, the estimator may well feel it prudent to price this item. The value of the making good of defects will depend on a number of factors, such as:

1. The design of the job and the choice and quality of the materials specified.
2. The quality of the workforce, the level of co-operation between the various trades and the standard of supervision on the part of the foreman, architect or clerk of works.
3. The extent to which domestic and nominated sub-contractors are used on the contract.

1. Where timber fascias and cladding are specified instead of, say, plastic or steel, it is likely that the maintenance costs will be higher due to perhaps shrinkage or warping taking place, paint peeling, etc. In general, where cheap, inferior quality materials are specified, higher maintenance costs can be expected than if good quality materials with a durable finish had been chosen. In addition, where cheaper materials are used, the element of doubt as to who should bear the cost of rectifying defective work increases. Thus, when floor tiles begin to lift, paint or varnish begins to peel from timber, or concrete tiles lose their colour – is it due to bad workmanship, fair wear and tear or poor quality materials originally specified? The doubt which arises can lead only to disputes and arguments between the architect and the contractor.
2. Where good quality work is specified, the capabilities of the workforce and the level of supervision by the agent or foreman must be of the highest standard. This may be particularly difficult to achieve if a bonus scheme is in operation, encouraging the men to increase their output as much as possible. Inevitably, as the men attempt to work at a faster rate, not only will there be a tendency to increase the amount of wastage of materials but the standard of workmanship could easily decline, resulting in more defects to be rectified later on.
3. Although the main contractor is responsible for *all* defective work, sub-contractors, both nominated and domestic, will, under the terms of their sub-contract agreements with the main contractor, be required to make good their own defects. Thus the higher the proportion of work carried out by sub-contractors, the

lower the maintenance costs will be which fall directly on the main contractor.

However, it should be borne in mind that although the main contractor will not be involved directly in the cost of making good defective sub-contractors' work, he still has to organise and co-ordinate each trade to carry out each item of work and to this end he will be involved in extra administration costs persuading his sometimes 'reluctant' sub-contractors to return to site to rectify their defects. (These administration costs can be offset to some extent by any discounts allowed to him or negotiated with each specialist.) As an added incentive to sub-contractors to effect an early completion of making good defects, there will of course be the promise of the release of the remainder of retention monies held on the issue of the certificate of completion of making good defects.

In order to be able to evaluate the effect of this clause, the estimator could assess (from his own experience and from reliable feedback from previous similar jobs) the type of work which generally needs attention.

A typical list may be as follows:

(a) Taps and valves leaking.
(b) Floor tiles lifting.
(c) Locks not working properly, door closers too stiff in action, requiring oiling or adjusting.
(d) Doors to be eased and adjusted.
(e) Warping, undue shrinkage in joinery work generally.
(f) Electrical faults, blown fuses, etc.
(g) Cracked panes of glass.
(h) Plasterwork to be made good around switches, pipes, conduits, reveals to be plumbed up.
(i) Paintwork, defective or marked (quite often by the contractor's own men during the final tidying up). Paint splashes on tiles, floors, etc.
(j) General brickwork, such as pointing to flashings, cleaning down brickwork, mastic pointing to frames, making good around pipes, etc.
(k) Paving slabs mis-aligned, tarmac to be made good around manholes, gullies, kerbs, etc.
(l) All the other minor items of work which remained incomplete at handover, such as the odd coat hook and door sign to fix, a mirror to plug and screw to the bathroom wall, and that 'particular' style of toilet-roll holder which the client insisted upon having. The one which the contractor ordered 9 months ago and delivery of which was promised within three days – its arrival on site still being eagerly awaited!

The costs of which may be assessed as follows:

(a)	Plumber 1 day at (say) £6.00 per hour (all-in, including transport, expenses, overheads, etc.)	£48.00
	Materials (say) £50	50.00
(b)	Floor tiler ½ day at (say) £5.00 per hour	20.00
	Materials – floor tiles and adhesive (say)	20.00
(c)	Joiner 2 days at (say) £5.00 per hour	80.00
(d) (e) (l)	Materials (say)	40.00
(f) (g)	Sub-contractors' work	NIL
(h)	Plasterer 1½ days at £5.00 per hour	60.00
	Materials (say)	£10.00
(i)	Painter 2 days at £5.00 per hour	80.00
	Materials (say)	£15.00
(j) (k)	Bricklayer and labourer 3 days at £5.00 and 4.00 per hour respectively	216.00
	Materials (say)	50.00
	Hire of mixer and other sundry plant costs (say)	100.00
	TOTAL COST	£789.00

This amount, if required could be included in the preliminaries bill against Clause 17 in the 'schedule of clause headings' section.

There is another cost to be taken into account, and that is the cost to the contractor in the form of lost interest or bank charges, due to retention money still being held during this six-month period. It has been stated earlier that the usual defects liability period is 6 months, although, there is provision in the appendix to the Standard Form of Contract for any length of time to be inserted by the employer.

Obviously, the longer the defects liability period, the longer the contractor will have to wait for his final portion of retention money. It is likely that the value of this item will have been taken into account when determining the percentage addition for overheads (see earlier).

5.4 Clause 18 – Partial Possession by Employer

Clause 18, together with the sectional completion supplement, provides for the employer to take possession of part or parts of the building before the completion of the whole works has been effected. If partial possession is required by the employer with proposed dates for handover being determined beforehand, then such dates will be written into the contract which had been suitably adapted by the use of the above supplement. This information would be brought to the contractor's attention in the tender documents. Notwithstanding any predetermined dates, the employer and the contractor can by mutual agreement arrange to complete and hand over the building in phases. If the work has to be carried out in a particular sequence or partial possession required, the consequences relating to cost would have to be considered by the estimator as follows:

5.4.1 Phased planning

The contract may be for the construction of, say, four separate blocks of flats with each block being handed over complete at intervals stated in the appendix; or the erection of a school comprising a mathematics block and an arts and practical block, the mathematics block being required to be completed and handed over before the completion of the whole job. In both cases, it is likely that the construction programme will contain an element of inefficient working. For example:

1 Bricklayers, having completed their work on one phase of the job, cannot proceed with the next stage since handover of that part of the work is a condition precedent to possession of the next section being given. Where this occurs, the contractor will have to find his men work elsewhere on another contract until they can return and carry on with the next phase. Once labour has been assigned to another job it may not be so convenient simply to 'whisk' it back again when required. Lack of continuity always tends to disrupt the smooth running of any contract.
2 With regard to plant and scaffolding hired for the contract, it may not be possible to maintain continuous usage. Mixers may be finished with for the time being only to be required again later on. The contractor may then be faced with the choice of either leaving such plant etc. standing idle on site whilst still paying hire charges on it, or bearing extra transport costs carting plant and machinery to and fro from his yard or depot. Again, once plant has been moved to another contract, it may not be readily available when required again for the next phase.

These extra costs, needless to say, are extremely difficult to assess with any degree of accuracy by the estimator. However, examining the preliminary programme and consulting with the plant manager and contracts manager will assist in his calculations.

5.4.2 Working whilst part of the building is occupied

If part of the building, or even one separate block, is handed over and subsequently occupied by tenants, the general public, or even worse, by school children, the contractor will almost certainly experience the following problems on site.

1. Safety. Curious passers-by and children are drawn like magnets to the workings of a building site and seem to have an amazing capacity of finding danger in the most innocent looking situations. All tools, plant, equipment and materials must be kept well away from prospective explorers not only for their own safety but also to prevent damage and theft.

To this end, the contractor must provide lock-up storage sheds, warning notices, barriers, hoardings, walkways, etc., all of which can be priced under the relevant section in the 'Preliminaries'.

2. Restrictions on working hours. Once part of the building is occupied, there may be restrictions on working hours. For example, work may have to cease while hospital operations are being performed, examinations take place or services take place in the case of the erection of, say, a church. Such restrictions should be brought clearly to the estimator's attention in the tender documents so that he may make the necessary allowances, either in the all-in labour rate or in the form of a lump sum addition in the preliminaries.

3. Private cars. The constant movement and parking of private cars in inconvenient places in the immediate vicinity of the works can hamper deliveries of materials and the free movement of traffic around the site.

This obviously interferes with the efficient working of the site and the estimator could make due allowance for this, again by including a sum in the preliminaries.

4. Defects. That part of the building which is handed over will commence its defects liability period from the date of the issue of the certificate of completion for that particular part of the work. This means that the foreman may be continually pulling men off the work in progress to attend to the rectification of defects. It is not so easy to avoid the complaints and aggravation of new tenants, say, when the builder is only a stone's throw away on the site!

Other paragraphs of Clause 18 can be beneficial to the contactor (albeit marginal), namely:

1 On the handing over of the relevant part of the building, the amount of retention held on the value of that portion of the work is reduced by a half, thereby helping the contractor's cash flow. (The remainder of the retention, of course, being released after the issue of the 'Certificate of Making Good of Defects' for that part of the work).

2 Again, upon the handing over of part of the building, there will be a proportional reduction in the amount of liquidated damages which could be levied on the contractor for the non-

completion of the remainder of the work. (The reduction will be in the same ratio as the value of work handed over is to the total value of the contract.)

3 The contractor reduces the value of work insured under Clause 22A by the amount of the relevant part, the employer taking responsibility of the building once it is handed over. This may reduce the amount of his premiums. (This may well be negligible, but the contractor's insurance company or broker would advise him on this point.)

5.5 Clause 19A – Fair Wages

Clause 19A forms part of the local authorities edition only and does not appear in the private edition. This clause calls upon the contractor to observe and comply with the terms and conditions of employment as laid down in the 'National Working Rule Agreement' and to pay the amount of wages as negotiated between the wage-fixing bodies. This guarantees operatives of at least a minimum wage although it is probable that most contractors will pay wages in excess of this minimum in the way of 'spot' or earned bonuses. It is most unlikely that contractors will attempt to pay less than the minimum since they would be unable to attract any labour.

Failure to comply with this clause amounts to default on the part of the contractor and could result in the contract being determined by the employer under Clause 27.3.

As regards cost, the estimator will use the agreed minimum wages as the basis for the calculation of his all-in labour rate.

5.6 Clauses 20, 21 and 22 – Insurances

Contracting is a risky business at any time, and insurances inevitably play an important part in the building process, the contract being very specific as to the extent of cover required for any particular project. There is more than one form of insurance applicable on any contract, the cost of which in the form of insurance premiums being included either as a lump sum in the tender summary, incorporated in the all-in rate or as a part of the overhead charges. The contractor can usually leave the business of arranging his insurance safely in the hands of his broker or insurance company since this is a specialised and complex topic. The contractor will provide all the necessary information relating to the value and general scope of his operations and his insurers can then calculate the amount of the premium.

It is more than likely that the insurers will quote the contractor a premium based on a 'blanket policy' covering all eventualities for a full year, obviously the premium being renewed annually. In certain cases, however, where additional risks are involved over and above those normally anticipated, an extra premium may have to be quoted. The insurance company itself may have to call on the services of an 'expert' in a particular field – a surveyor or engineer, perhaps, to assess the degree of risk involved. Obviously, the amount of the premium will depend on the extent of such risks.

Having said the estimator can rely on his accountant and insurers to advise him on the amount to include within his tender, it is still necessary for him to fully understand the implications of these clauses, which are as follows:

5.6.1 Clause 20

This simply requires the contractor to indemnify (protect) the employer against any claim arising out of injury or death to any person or damage to property real or personal.

In the forms of sub-contract agreements a similar provision exists whereby the sub-contractor is required to indemnify the main contractor in respect of the same.

5.6.2 Clause 21.1

This specifically calls upon the contractor to take out insurances to cover his liability under the previous clause. The limit of indemnity for all third-party liability for any one occurrence or series of occurrences arising out of one event is to be stated in the appendix against 21.1.1. A typical amount would be in the region of £1,000,000 (Many contractors may have a greater cover than this, but the figure included in the bills is a minimum requirement.) The employer would have sought advice from his insurers in order to ascertain a realistic figure.

5.6.3 Clause 21.2

Where the employer requires a specific risk of damage to any property other than the works to be covered, the cost of insuring in such a case is dealt with by the inclusion of a provisional sum at tender stage. If, subsequently, the employer instructs the contractor to take out an insurance, the cost of the premium will be set against the provisional sum in the final account. Thus, unless a provisional sum is allowed in the bill of quantities, the estimator need make no provision for this item.

5.6.4 Clause 22

This refers to insurance of the works against fire etc.

The employer faces a choice under Clause 22 with sections not applicable being deleted. Clause 22 is made up as follows:

22A – applicable to the erection of a new building if the contractor is to insure.

22B – applicable to the erection of a new building if the employer is to bear the risk.
22C – applicable to alterations of or extensions to an existing building. In this case there is no option – the risk to be borne by the employer.

The part of this clause which is to apply will be brought to the estimator's attention in the bill of quantities. The two remaining sub-clauses would be deleted in the form of contract.

If, therefore, on a number of contracts throughout the year the employer bears the risk under Clause 22, the amount of the contractor's premiums may be reduced accordingly. His insurer's advice would be obtained on the matter.

5.7 Clause 23 – Possession and Completion

The actual dates for possession of the site and completion of the works, determined by the employer. should be brought to the estimator's attention in the tender documents and prior to that in the preliminary enquiry for invitation to tender. These dates would subsequently be written into the appendix and form part of the conditions of contract.

These dates are of utmost importance since they determine:

1 The Contract Period; and
2 The time of year for starting and finishing the work.

It is important that the **actual** dates are determined before the tender documents are sent out in order that all contractors are pricing on the same basis. Vague statements such as 'date for possession – to be agreed' leaves one of the most crucial aspects of the pricing of the job open to individual interpretation and should be avoided. This is particulary important where a 'firm price' tender is required since the estimator must allow not only for all anticipated increases during the actual contract period, but also for all increases occurring between the date of tender and the starting date on site. If this starting date is not stated, how long should be allowed?

There are occasions when this period of 'limbo' can be as much as six months or more. Whether this delay is due to the employer's inability to gain possession of the site or to his being unable to obtain finance is of little interest to the contractor. When work is hard to come by, there will be considerable pressure upon the contractor to extend the period of his original firm price offer, often with no promise of additional reimbursement. When an actual date for possession is inserted, even if it is six months after the date for the return of tenders, at least the estimator can allow for this within his price.

The 'Code of Procedure for Single Stage Selective Tendering 1977' gives little guidance on the time lapse between submission of tenders and the starting date on site. Under Clause 8, it recommends that this period should be neither too short nor too long. The code does suggest, however (*inter alia*), that tenderers should be required to make offers based on the same period of completion. This limits the basis of competition to price only.

The effect upon the tender of (1) above (the *Contract Period*) will be considerable, since this will affect:

1 The percentage addition to be included for inflation on firm price contracts.
2 A large portion of the Preliminaries bill. The value of the following preliminary items being directly time-related.

 (a) Person in charge;
 (b) hire of storage and site accommodation;
 (c) electricity, telephone charges;
 (d) daily travelling and fare allowances;
 (e) plant hire;
 (f) transport charges.

The following considerations relating to (2) above (*the time of year for starting and finishing the work*) would have to be taken into account.

1 The starting and finishing dates will determine the amount of winter working. If the contract is in its early stages during the winter months there will be more non-productive time due to stoppages for bad weather. The level of efficiency of operatives and machinery working in muddy, waterlogged conditions will fall dramatically; the excavation rates must obviously be based on a 'swings and roundabouts' method since no one could possibly predict weather conditions.
2 The cost of keeping roads and footpaths clean and free from mud would also be much higher if the excavations were carried out in winter.
3 The cost of electricity for site lighting would be more if the 'shell' of the building was being constructed in winter with no permanent electricity supply yet available.
4 The cost of keeping excavations free from water (not spring or running water) could be considerable during winter. The cost of pumping or dewatering being incorporated into the excavation rates or priced separately in preliminaries.
5 The finishing date may well determine to a large extent the cost of drying out the building. Whether the contractor is able to use the building's own heating system or whether he chooses to use jet air blowers or similar, the cost of fuel and attendance will always be much more during the cold wet days of winter. If the completion date is in the summer months it may be that the building will dry out naturally with the contractor being involved in little or no cost at all.

5.8 Clause 24 – Damages for Non-completion

Damages for non-completion or 'liquidated damages' is an amount of money which is intended to be deducted by the employer from money outstanding to the contractor should he fail to complete the works by the stated date for completion under Clause 23.

The contractor may, however, be relieved of some or all of the effects of this clause if he obtains from the architect a valid certificate(s) of extension of time issued under any of the twelve sub-clauses of Clause 25.

The amount of money and the rate at which it is to be deducted is written into the appendix and is brought to the estimator's attention in the tender documents.

Often the amount of liquidated damages is nominal, but where the employer could be involved in large losses if the building is not completed on time (e.g. loss of rent or sales, etc.) the figure could be substantial and if this is the case the estimator must consider the following points:

1 If the amount of the liquidated damages is substantial, it could rebound on the employer by discouraging contractors from pricing altogether, particularly if the contract period seems 'tight' anyway. Alternatively, if contractors do submit prices, the tenders may contain a premium of perhaps three or four weeks' liquidated damages to cover the contractor in the event of his exceeding the contract period.
2 The estimator, together with perhaps the plant manager and contracts manager, will be encouraged to plan the execution of the job very carefully in order to avoid liquidated damages. This would include preparing a preliminary programme and method statement showing the allocation of time to each stage in order to complete the whole of the work by the due date for completion.

   The outcome of such a meeting may be the proposed intensive use of mechanical plant, the cost of which will be reflected in the tender.

It is, of course, in the contractor's own interest to complete the work within the contract period regardless of liquidated damages. Except in the case of (1) above, therefore, there would be no need for this clause to affect the tender.

5.9 Clauses 25.4.10.1 and 25.4.10.2 – Extension of Time – Availability of Labour, Goods and Materials

Clause 25 provides for the architect to be able to grant the contractor an extension of time if he is delayed by any of the occurrences listed under this heading. Sub-clause 25.4.10 is particulary important since it is frequently deleted from the contract by the employer. In the 1963 edition, of the JCT form, a footnote drew employers' attention to this sub-clause, and the possibility of its deletion, although in the 1980 edition, this footnote has disappeared.

This particular sub-clause allows the contractor to claim an extension of time if he is, for reasons beyond his control, unable to obtain (i) the necessary labour and (ii) the necessary goods and materials with which to complete the contract. This, therefore, removes a great deal of the element of risk from the contractor, the employer accepting the costs to himself of any delays resulting from the effect of this sub-clause.

However, as previously stated, this part of Clause 25 may be deleted thereby shifting the responsibility and the risk of being able to obtain the required labour and materials necessary back on to the contractor. It is reasonable to expect that, by accepting such a risk, the contractor should want to include a compensating amount within his tender, the employer in effect therefore paying for the deletion of this part of the contract.

Since deletion can affect both the price and number of tenders submitted, the employer should consider carefully whether or not it would be in his own interests to delete or to leave the whole clause intact. Market conditions are constantly changing, the availability of materials fluctuating widely depending to a large extent on the economic climate at any particular time. The extent to which the risk of shortages of labour and/or materials should be placed upon the contractor must be judged in each individual case depending on the conditions prevailing at the time. In normal times, when a plentiful supply of labour and materials can be expected, this sub-clause could reasonably be deleted in full, providing the architect in turn tries to ensure that all materials specified are readily obtainable.

The estimator, if having determined that 25.4.10.1 and 25.4.10.2 have been deleted, may wish to take the following items into account when preparing the tender:

1 Since no extension of time will be granted if he is unable to secure the necessary labour with which to complete the works (25.4.10.1), the contractor may well have to pay over the odds in order to obtain the 'quantity' of labour needed in the time allowed. Such extra labour costs may comprise:

   (a) The payment of overtime rates to his men in order to be able to work beyond normal working hours and at weekends.
   (b) The payment of incentive bonuses in order to attract labour from elsewhere and to increase productivity.
   (c) Additional transport and travelling costs and perhaps lodging allowances in connection with the importation of labour from outside the district.

(The costs of (a) and (b) above could be incorporated into the 'all-in' labour rate, the value of (c) assessed separately and included in the preliminaries bill under 'travelling, transport etc.')

2 With the deletion of 25.4.10.2, No extension of time can be granted if the late delivery or non-availability of materials contribute to delays on the contract. (In the latter case, it would be reasonable to expect that the architect will specify an alternative material.) Extended delivery of materials, however, can pose considerable problems to the contractor, especially since delivery periods stated on quotations often fall short of the actual supply dates!

 The contractor, in such circumstances, could find himself involved in additional administrative costs 'chasing up' deliveries, together with the extra costs of perhaps being forced to obtain materials from other suppliers or merchants outside his normal area of working. He may not be able to obtain the same favourable trading terms from such suppliers and may even have to arrange collection and delivery of the goods himself.

(An additional percentage or premium may be added to the price of materials which the estimator feels he will have particular difficulty in obtaining. The cost then being incorporated into the unit rates and distributed throughout the bill within each affected trade.)

3 Should labour and materials be particularly difficult to secure, and if the amount of liquidated damages is substantial, the estimator may decide it prudent to build into his tender perhaps three or four weeks' liquidated damages as a 'hedge' against non-completion (see Clause 24).

 The inclusion of such an amount does have its drawbacks, of course – the more the estimator 'loads' up his tender in this way, the less competitive his price becomes. In many cases, when competition is fierce, the successful tenderer is often the one who makes the most mistakes or who has failed to make adequate provision within his tender for such likely occurrences as stated in this sub-clause. A pointer perhaps to the need for an overhaul of the present tendering system?

5.10 Clause 30 – Certificates and Payments, Retention

There will be stated in the Appendix to the contract, against Clause 30, the following information relating to certificates and payments:

1 The period of interim certificates.
2 The retention percentage.
3 The period of final measurement and valuation.

5.10.1 The period of interim certificates

Although usually monthly, the period can be any length of time the employer chooses – fortnightly or perhaps even every six weeks. The contractor must finance the work himself until a payment is received and, since his men obviously require paying weekly, the longer he has to wait for each payment, the higher will be the cost to him of financing the project (probably in form of additional bank charges).

For a contractor with several large contracts in progress all at the same time, the cost of financing his whole operation will be considerable. It is therefore vital that a steady, even cash flow is maintained with payments being made at as short an interval as possible. Such costs may be left to the accounts department to work out, the resulting amount forming part of the contractor's 'overheads'.

5.10.2 The retention percentage

The amount of money outstanding, in the form of retention, affects the contractor's cash flow and adds to the cost of financing the project. Retention is an inevitable feature of contracting, the purpose being to hold back a certain proportion of the value of completed work in order to protect the employer against default by the contractor.

The employer pays for this requirement indirectly since an element of the overhead charges will relate to the cost to the contractor of having this money outstanding. Again the cost of bank charges or loss of interest is an exercise for the accountant, who can look back at previous years' trading figures to establish, on average, what proportion of the 'turnover' is held as retention. Retention is normally 5 per cent of the value of the work with half of this being released on practical completion. The contract allows for a lower rate than 5 per cent to be applicable, but this is at the employer's discretion. It is unlikely, however, that any saving in the cost of financing retention due to lower percentage being inserted in the contract will have a significant effect upon the tender.

5.10.3 Final measurement and valuation

Although the stated period for the final measurement is six months, this is often delayed due to pressure of work on current contracts elsewhere, the settlement of claims the production of the final fluctuation claim, etc. As before, the longer the final settlement drags on, the longer the contractor will have money outstanding, again reflected in bank charges not budgeted for.

5.11 Clauses 38, 39 and 40 – Firm Price/Fluctuating Price

Clauses 38, 39 and 40 without doubt have the greatest single bearing upon the tender figure than any other clauses in the contract and go to the very root of the basis of the calculation of unit rates and preliminaries.

Since these are such fundamental clauses within the contract, they must be thoroughly understood by the estimator who must recognise immediately whether the employer requires a 'firm' or 'fluctuating' price for the work.

Often, the estimator will be able to discover which clause is to apply by examining the appendices to the bill of quantities. (Usually these can comprise three or four pages inserted immediately before the preliminaries bill, or could even form part of the preliminaries bill itself.)

The following is a typical extract from a bill of quantities, showing how the information relating to these clauses may be given.

Roofhead Housing Association
erection of 25 bungalows at
Jamestone Road, Southtown   Contract No. 376

Appendix 2
Summary of alterations and deletions

Insurance of Works	Clause 22(A)	Will be deleted.
Against Fire etc.	Clause 22(B)	Will be deleted.
Extension of Time.	Clause 25.4.10	Will be deleted.
Fluctuations	Clause 38 Clause 40	Will be deleted.

Appendix to conditions of contract

The Appendix to Conditions of Contract shall be completed by the insertion of the following:-

Insurance Cover for Any One Occurrence or Series of Occurrences Arising Out of One Event.	Clause 21.1	£1,000,000
Percentage to Cover Professional Fees.	Clause 22A	Nil
Defects Liability Period.	Clause 17, 18 and 30	Six months.
Date for Possession.	Clause 23	1st July, 1983
Date for Completion.	Clause 23	30th June, 1984
Liquidated and Ascertained Damages.	Clause 24	At the rate of £100 per week
Period of Interim Certificates.	Clause 30.1.3	One Month
Retention Percentage.	Clause 30.4.1.1	Five per cent
Period of Final Measurement and Valuation.	Clause 30.6.1.2	Six months.
Percentage Addition.	Clause 38.7 or 39.8	Five per cent.

From the above information it will be clear to the estimator that a fluctuation price is required, with a 5 per cent addition on top of the net amount of labour and material fluctuation being paid to the contractor during the course of the contract, as claims are submitted and agreed.

Since Clauses 38, 39, 40 are of such importance, it may be as well to devote some space to analysing them in detail before going on to actually calculate the effects they have upon the tender figure. Contracts are let on either a 'firm price basis' (usually those where the contract period is less than twelve months) or a 'fluctuating price basis', there being a choice of method of reçovering increases, i.e. the 'traditional/conventional' method or the 'formula' method.

Firm Price	**Fluctuating Price**		
Clause 38 Applies (Clauses 39, 40 deleted)	*Formula method*	OR	*Traditional method*
	Clause 40 applies (Clauses 38, 39 deleted)		Clause 39 applies (Clauses 38, 40 deleted)

The fluctuation price clauses are published as separate documents to the JCT Standard Form of Contract and are produced in different versions to be read in conjunction with:

1 Private edition with quantities.
2 Private edition without quantities.
3 Local authorities with quantities.
4 Local authorities without quantities.

Two further variations of the private and local authorities editions are produced for use with approximate quantities.

One *only* of the following three clauses may be used in conjunction with the main contract, depending on the client's requirements in connection with fluctuations:

	CLAUSE 38	Firm price clause.
or	CLAUSE 39	Fluctuation price clause using the 'conventional' method of recovery of increased costs.
or	CLAUSE 40	Fluctuation price clause using the NEDO formula method of recovery of increased costs.

The two clauses which are not applicable would be deleted, with the contractor's attention being drawn to the amendments to the standard clauses in the Preliminaries.

5.11.1 Clause 38

Used mainly for contracts lasting less than twelve months where a client can reasonably expect tenderers to anticipate inflation levels and thereby commit themselves to a 'firm' price by including all allowances for increased costs within their tenders. However, although Clause 38 is normally looked upon as the 'fixed' price clause, *certain* categories of increased costs are still allowed to the contractor – these relate to:

1 Increases/decreases in contributions, levies and taxes in connection with the employment of labour – Clause 38.1.
2 Increases/decreases in contributions, levies or taxes in connection with the purchase of materials, goods, electricity or (where specifically stated in the bills of quantities), fuels – Clause 38.2.

Increases/decreases are paid to or allowed by the contractor in accordance with any changes which may take place in the above rates of contributions, etc. after the date of tender irrespective of whether they were known about at the time of pricing the work.

Domestic sub-contractors
Where Clause 38 is applicable to the main contract, similar provisions are to be incorporated into any form of sub-contract agreement where the contractor chooses to sub-let portions of the work – Clause 38.3

Other provisions of Clause 38
Clause 38.4 relates to the practical applications of Clause 38 generally as follows:

Clause 38.4.1	The contractor is to give a written notice to the architect informing him of any changes in the rates of contributions, etc. described above.
Clause 38.4.2	Such notices are to be made within a 'reasonable' time and are to be regarded as a condition precedent to any payment being made.
Clause 38.4.3	QS and contractor to agree the amounts of fluctuations involved.
Clause 38.4.4	Any amounts involved to be added to or deducted from the contract sum and are not subject to retention.
Clause 38.4.5	Contractor to submit calculations to support his claim.
Clause 38.4.6	Amounts paid/deducted in respect of fluctuations to be net with no effect on the amount of profit already contained in the contract sum.
Clause 38.4.7	No fluctuations to be allowed for increases/decreases taking place after the date for completion (unless an extension of time has been granted).
Clause 38.4.8	Lists the conditions which must be applicable for Clause 38.4.7 to apply.

Work to which fluctuations are not applicable
Certain types of work are excluded from any fluctuation calculations – these are set out in Clause 38.5.

Definitions

A number of important definitions in connection with the practical aspects of this clause are dealt with under Clause 38.6 – these relate to:

1 'Date of Tender'.
2 'Materials and goods'.
3 'Workpeople'.
4 'Wage fixing body'.

Percentage Addition

Clause 38.7 provides for a percentage addition to be applied to the net amount of the fluctuations calculated in accordance with the above conditions. The actual figure varies from one contract to another, depending on how much a client is prepared to pay, and can range from 0 to say 20 per cent. The client must decide upon the percentage addition he considers suitable when the contract documents are being prepared so the amount can be stated in the appendix as required by this clause. The purpose of the percentage addition is to reimburse the contractor to some degree for loss of overheads and profit on the amount of the increased costs and for costs incurred in the administrative work involved in the preparation of the claims.

5.11.2 Clause 39

Used where the contractor is allowed to recover increased costs as the contract progresses; with increases being calculated on the basis of the net difference between prices of materials and rates of wages ruling at the date of tender and those in operation when materials are subsequently purchased and labour used on the project. Although Clause 39 is a 'fluctuation' clause, only certain categories of increased costs are recoverable, leaving the contractor with a 'shortfall' which must either be accepted or included in the tender figure as a lump sum adjustment.

The areas of increased costs which **may** be recoverable are as follows:

1 Changes in the rates of wages and other payments made to 'workpeople' together with corresponding increases in respect of other persons employed on the works (e.g. site clerk, storeman, QS, agent, etc.) even though their wages/salaries may not have actually increased. In addition, the corresponding increases in NI and any other contributions etc. made by the contractor in response to increases in wages and allowances will also be paid – Clause 39.1.
2 Changes in the rates of contributions, taxes, levies, etc. in respect of the employment of labour or the purchase of materials and fuels as Clause 38.

3 Changes in the market price of materials, goods and electricity and, if stated in the BQ, fuels. Providing:
   (a) The materials and fuels have been specified on a 'basic price list' submitted by the contractor with his tender and which shows the prices of materials which have been used in the compilation of the tender.
   (b) The materials are purchased from the same suppliers as named on the basic price list.
   (c) The materials are purchased in the same units as those stated on the basic price list.

Further provisions relating to domestic sub-contractors, written notices, work not applicable, definitions, percentage addition, etc. are all as for Clause 38.

5.11.3 Clause 40

Used when fluctuations are allowed, but where the method of calculating the amount of the fluctuations is based on the movement of index numbers. Clause 40 introduces a further document into the contractual arrangements by referring to the 'Formula Rules' which are to be used in the practical application of this clause. Using this method of recovery of increased costs, **all** of the contractor's costs are subject to adjustment since all the items within the BQ are allocated to 'work categories' or 'work groups', the value of which is subsequently adjusted by means of 'the formula'. It follows that since *all* of the items within the BQ are subject to adjustment, the contractor is able to recover increases in overheads, profit and preliminary items which were excluded under Clause 39 and any 'shortfall' virtually eliminated.

Adjustments in respect of Clause 39

It has been calculated that the 'non-recoverable' increased costs which obviously vary depending on the nature of the contract can form as much as 40 per cent of the total amount of increases actually incurred by the contractor. The difference between the total increases incurred and the amount of the increases recoverable under the contract is commonly referred to as the shortfall. This shortfall can be mitigated to a small extent by a percentage being inserted against 39.8 as previously described.

Very simply, the effects of 39.8 may be illustrated as follows:

Total *net* amount of increased costs incurred	£20,000
Total amount recoverable under the contract (say 60%)	£12,000
Therefore amount of increases *not* recoverable	£ 8,000
Add Clause 39.8 (say 10%) to net amount of recoverable increases (10% of £12,000)	£ 1,200

Thus it can be seen that the amount of non-recoverable increases (£8,000) is offset to some degree by the £1,200 recovered under Clause 39.8, but it still falls far short of what the contractor would want to recover.

For a long time after fluctuation clauses became common, many contractors were unaware of the magnitude of this shortfall, thereby incurring in some instances heavy losses without realising why.

To make matters worse, in practice many employers have chosen to insert 'Nil' against the percentage addition for 39.8, thus in effect putting the contractor back to square one. Happily, however, even this extreme condition can be put right.

Table 5.1 is an example of the calculation an estimator may make on a typical contract, regardless of the percentage for 39.8, in order to make full allowance within his tender of any anticipated shortfall.

Let us assume that the contract is for the erection of 25 bungalows and the contract particulars are those listed on page 153, i.e. twelve month contract period let on Clause 39 with the percentage addition against Clause 39.8 being 5 per cent.

Assume the estimator's preliminary tender figure is made up as shown in Table 5.1.

Table 5.1

	(£)
Preliminaries	40,000
Excavation (including £10,000 sub-let to groundworks sub-contractor)	20,000
Concrete work	30,000
Asphalt work (all sub-let)	14,000
Brickwork and blockwork	44,000
Woodwork	64,000
Metalwork	4,000
Plumbing (all sub-let)	27,000
Glazing (all sub-let)	10,000
Painting	12,000
External works (including £16,000 sub-let to tarmacadam sub-contractor)	32,000
PC and provisional sums	124,000
Attendances	3,000
Insurances	6,000
Bond	2,520
Water	1,200
Preliminary tender figure (excluding overheads and profit)	£433,720

Let us further assume that this particular contractor's pricing policy is to attempt to recover 10 per cent overheads and 5 per cent profit on the *total value* of his own work alone. (Perhaps he wants to keep his tender competitive and is quite happy not to price overheads and profit on his own private sub-contractor's work, but includes the net amount of each quotation within his tender, taking advantage only, perhaps, of any discounts offered.)

The first step in the calculation would be to extract the value of his own work from the above preliminary tender figure, as shown in Table 5.2.

Table 5.2

	(£)
Preliminaries	40,000
Excavation (£20,000–£10,000)	10,000
Concrete work	30,000
Asphalt work	nil
Brickwork and blockwork	44,000
Woodwork	64,000
Metalwork	4,000
Plumbing	nil
Glazing	nil
Painting	12,000
External works (£32,000–£16,000)	16,000
Attendances	3,000
Total value of contractor's *own* work	£223,000

(The amounts for the bond, water and insurances could be included in the above figure, but in this instance it is assumed that the contractor wishes to include the net amounts in his tender.)

The estimator must then anticipate all the increased costs which could occur within this sector of the work, which means assuming a figure for inflation. It is unlikely that the labour and material costs will increase by the same percentage throughout the contract, so let us assume that, on average, throughout the 12 month period of the contract, labour costs will increase by, say, 8 per cent, material costs will increase by say, 15 per cent, and that plant charges, supervision, transport, overhead charges, etc. will increase by 10 per cent. (Remember it is the work in the last category upon which the contractor is *not* entitled to recover any increases.)

Having assumed such inflation figures, the estimator must then break the value of his own work down further into labour, materials and plant, etc. See, for example, Table 5.3.

Table 5.3

Trade	Labour material split (%)	Value of labour (£)	Value of materials (£)	Value of plant
Excavation	80–20	8,000	2,000	In preliminaries
Concrete work	60–40	18,000	12,000	In preliminaries
Brickwork	45–55	19,800	24,200	In preliminaries
Woodwork	50–50	32,000	32,000	In preliminaries
Metalwork	40–60	1,600	2,400	In preliminaries
Painting	70–30	8,400	3,600	In preliminaries
Ext. works	60–40	9,600	6,400	In preliminaries
Attendances	50–50	1,500	1,500	In preliminaries
		£98,900	£84,100	

(For ease of calculation, it has been assumed that all plant, etc. has been priced in the preliminaries section, although this may be equally included within the unit rates in each trade. The principle. however, remains the same.)

In order to be more accurate, the estimator could assume a different inflation figure for each trade instead of applying a blanket percentage covering the whole of his own work. Again, a single figure has been chosen so as not to complicate what is essentially a straightforward calculation.

In practice, of course, the estimator does not have to assume a labour/material split within each trade, as he will price the labour and material elements of each item separately thereby obtaining page totals and subsequently trade totals for each.

1 *Labour* – during the course of the contract it is anticipated that labour costs will increase by:

   £98,900 × 8% = £7,912

2 *Materials* – during the course of the contract it is anticipated that material costs will increase by:

   £84,100 × 15% = £12,615

Preliminaries. Since the plant costs have been priced in the preliminaries section, the bulk of the £40,000 will fall into this and the third category of work (i.e. supervision, transport, hire of site huts, etc.) upon which no increased costs can be claimed. Perhaps, therefore, only a relatively small proportion of the £20,000 total will fall into the categories of 'pure labour' and 'pure materials'.

Table 5.4

Category of work	**Amount recoverable under the terms of the contract**	**Amount the contractor wishes to recover**
(A) Work trades		
(i) Labour	£7,912 in increases + 5% in respect of 39.8	£7,912 in increases + 10% overheads + 5% profit
	= £7,912 + £396	= £7,912 + £791 + £396
	= £8,308	= £9,099
(ii) Materials	£12,615 in increases + 5% in respect of 39.8	£12,615 in increases + 10% overheads + 5% profit
	= £12,615 + £631	= £12,615 + £1,262 + £631
	= £13,246	= £14,508
(B) Preliminaries		
(i) Labour	£960 in increases + 5% in respect of 39.8	£960 in increases + 10% overheads + 5% profit
	= £960 + £48 = £1008	£960 + £96 + £48 = £1,104
(ii) Materials	£600 in increases + 5% in respect of 39.8	£600 in increases + 10% overheads + 5% profit
	£600 + £30 = £630	£600 + £60 + £30 = £690
(iii) Plant, overheads, transport, etc.	Nil (no increases allowed)	£24,000 in increases + 10% overheads + 5% profit
		£2,400 + £240 + £120 = £2,760
	£23,192	£28,161

Assume, therefore, that the £40,000 figure for preliminaries comprises: £12,000 labour costs, £4,000 material costs, with the remainder, £24,000, being made up of items upon which increases cannot be recovered.

Applying the same anticipated inflation figures as before, labour costs will increase by:

£12,000 × 8% = £960

material costs will increase by:

£4,000 × 15% = £600

and, assuming an inflation figure of 10 per cent for the remainder of preliminaries, the cost of such items will increase by:

£24,000 × 10% = £2,400

The above results can be tabulated as shown in Table 5.4. From this, it can clearly be seen that on this particular project the contractor would want to recover £28,161, yet under the terms of the contract he will be able to recover only £23,192. Thus, he will be involved in a shortfall of £28,161 – £23,192 = £4,969, and if he wishes to obtain a 100 per cent recovery of fluctuations, he must include this figure as a 'firm price element' within his tender.

This adjustment, together with a percentage addition for overheads and profit based on the net value of his own work previously calculated, will comprise the contractor's final tender figure.

(It is a matter of choice for the contractor whether the 10 per cent overheads and then the 5 per cent profit are based on the net cost as in the example, or whether the 5 per cent profit margin is based on the aggregated net cost plus overheads.)

The advantages and disadvantages of the use of Clause 39 when recovering fluctuations can be summarised as follows:

Advantages

1 **Actual** increases in labour and material costs are recoverable since fluctuation calculations are based on the contractor's own labour returns and his invoices for materials purchased.
2 Simple to operate.
3 Capable of being audited accurately (important particularly for work executed by local authorities and other public bodies, where accountability is essential).

Disadvantages

1 It is a costly, time-consuming operation.

2 The contractor is not able to recover *all* of his increased costs. He is entitled to recover only:
 (a) Increases in wages (but not incentive bonuses, etc.).
 (b) Increases in the market prices of materials, **provided**:
  (i) The contractor purchases the materials from the same supplier as that stated on the basic price list; see Fig. 5.1.
  (ii) The materials are stated on the basic price list.
  (iii) The materials are purchased in the same quantities as those stated on the basic price list.
 (c) Increase in the cost of labour and materials, due exclusively to changes in the rates of taxation, levies, duty, etc. brought about by Government legislation.
 (Notwithstanding the above, the estimator is still able to make an allowance for these items in his tender as the foregoing calculation shows.)

3 There will always be a time lag between the contractor actually being involved in increased costs and his being reimbursed for them. (His wage bill will be a weekly cost, yet at best he can only recover his labour increases in arrears on a monthly basis, i.e. at the time of each interim valuation.) This obviously affects his cash flow.

Notes:

(a) The basic price list, although required to accompany the tender, is frequently submitted at a later date if the contractor's tender is under consideration and his priced bill of quantities is requested. This is because many prices will have been obtained by telephone and written confirmation from merchants and suppliers will not be received by the date tenders are due to be submitted. Furthermore, the production of the basic price list, together with all supporting documents (quotations, price lists, etc.) is a time-consuming operation and one which the estimator will only want to complete if his tender looks favourable.

 At tender stage, the estimator will probably return the basic price list not filled in but endorsed with the words 'information will be submitted at a later date if requested'.

(b) Materials omitted from the basic price list, even accidentally, will not qualify for adjustment for fluctuations. It is essential, therefore, that the estimator carefully checks the list to make sure that all materials upon which the contractor will subsequently want to claim increases are included. Materials of a minor nature may be deliberately excluded from the basic price list, with their corresponding anticipated increased costs being included in the pricing of the work on a 'firm price' basis.

ROOFHEAD HOUSING ASSOCIATION
ERECTION OF 25 BUNGALOWS
SOUTHTOWN

CONTRACT NUMBER 376

Appendix 4

BASIC PRICE LIST – CLAUSE 39.3.1

The contractor is required to enter below a list of materials or goods, together with the name of the supplier, which form a substantial amount in the contract sum and upon which he or his subcontractors have based their Tender.
The basic prices against each item are to be 'delivered to site' and adjustments to this list will be made in accordance with Clause 39 of the conditions of contract.

The contractor will be required to produce quotations to substantiate any basic prices listed below.

SUPPLIER	MATERIALS OR GOODS	UNIT	BASIC PRICE
Northern Manufacturing Ltd.	Steel lintols	As price list	dated 30.3.82
Southtown Timber Engineering	Roof trusses	Each	30.00
Clayton Brick Co. Limited	65 mm common bricks	1000	91.00
	Moorland rustic facing bricks	1000	125.50
	Class 'B' Engineering bricks	1000	142.75
S. Greenland (Roofing) Ltd.	Concrete tiles	1000	308.00
	120° Ridge	100	81.00
	Verge clips	100	25.00
	BS.747 felt	20 m Roll	13.90
	38 × 25 mm tanalized laths	100 m	21.20
	Std. clips	100	5.00
Acemix Concrete Limited	R.M.C.	As price list	dated 29.3.82
Precast Concrete Limited	Paving slabs, seats, bins, kerbs etc.	As price list	dated 1.4.82
Wicker Goods Limited	Door mats	Each	15.00
West Stone & Terrazzo Limited	Padstones, lintols	As price list	dated 28.3.82
Alliblock Limited	Concrete blocks	As price list	dated 29.3.82
Builders Supply Limited	500 gauge visqueen	200 m Roll	23.10
	P.V.C. damp course	m^2	85p
	Cement	Tonne	66.00
	Drainage items	As price list	dated 30.3.82
	Wallboard sheet	100 m^2	150.00
Halls Quarry Limited	Sand	Tonne	10.80
	Gravel	Tonne	11.20
Southern Asphalt Limited	Bitumen	Drum	60.00
	19 mm aggregate – Tarmac	Tonne	29.10
	10 mm ditto	Tonne	33.50
James Smith Timber	Carcassing timbers, joinery timbers, skirtings, architraves etc.	As price list	dated 27.3.82
Rainbow Paint Company	Primer, undercoat, gloss, emulsion etc.	As price list	dated 28.3.82

Contractor's Signature ______________________

Address ______________________

____

____

Date ______________________

Fig. 5.1

Adjustments in respect of Clause 40

In a generally conservative industry, perhaps the biggest single innovation in recent years has been the introduction of a 'formula' method of reimbursing contractors with increases in cost when a fluctuating price is required. This section is intended to give a brief introduction into the workings of the formula and how it affects the estimator. There are many excellent reports and publications available on this subject, and the reader is advised to study further in order to obtain a comprehensive working knowledge of the formula both in pre- and post-contract practice.

Although it may seem obvious, those new to the topic of fluctuations are reminded that the formula method is not intended to replace the conventional system of recovering increases, but offers an alternative, and employers must choose which method they feel will be the most suitable, given their own particular circumstances.

The formula has been incorporated into both the private and local authorities editions of the JCT form of contract by the introduction in 1975 of a new clause – 31F, now Clause 40 in the 1980 edition. This brought within the scope of the contract a further document (standard form of building contract – *Formula Rules*, reference being made to this in Clause 40 itself and in the appendix). This implies that any revision to the *Formula Rules* automatically revises the conditions of the main contract; it is vital, therefore, that contractors keep abreast of all latest developments.

Having examined the recovery of fluctuations by the operation of Clause 39, its weak points are apparent. Although it is simple in principle and is capable, albeit laboriously, of being audited accurately, the system embodies a disturbing number of disadvantages. It was decided, therefore, that in the light of these disadvantages an attempt should be made to find another method of recovering increases where a fluctuating price was required. A study by the National Economic Development Office was subsequently conducted in 1969 into the problem and a report was produced some years later suggesting the possibility of the use of a 'formula' method for adjusting cost fluctuations.

The advantages of such a system were put forward as being:

1 Administration costs would be reduced dramatically, since contractors would not have to produce fully priced out time sheets and copies of invoices for materials purchased in support of their fluctuation claims.
2 Payments for increases would be made quickly at the time of each valuation, the contractor thus receiving his money much quicker compared with the conventional method.
3 Any problems in defining market price and in the purchasing of materials from particular suppliers in particular quantities are eliminated, since all information required to operate the system is

based on Government statistics which are published monthly.

4 Perhaps the most important of all the improvements over clause 39 is the fact that the contractor is able to obtain a far greater recovery in increased costs with the formula than with its conventional counterpart. Whereas, by using clause 39 the contractor is reimbursed for labour and materials only, under the formula method he recovers increases in all cost categories, including overheads, profit and preliminaries. The 'shortfall' element mentioned previously is all but eliminated.

Very briefly, the 'Formula' is a method of adjusting the contract sum at each interim valuation by adjusting the value of work included in that valuation in accordance with the movement of the index numbers published in the monthly bulletins.

5.11.4 Sub-contractors

As under Clauses 38 and 39, the contractor's own private sub-contractors would be expected to price their section of specialist work in accordance with the provisions of clause 40. The estimator must check, therefore, that all quotations received from prospective sub-contractors have been based on the formula method of recovering increases, the terms and conditions of the main contract being embodied in the form of sub-contract agreement.

Clause 40 and the 'Formula Rules' do not apply to nominated sub-contractors, however. Fluctuations on their work are calculated in accordance with conditions previously agreed between the architect and the sub-contractor.

(Increases being adjusted using specialist formulae or by reference to the relevant work category index numbers. This provision relating to nominated sub-contractors does not, of course, affect the pricing of the bill of quantities at tender stage, since a PC sum is used to cover the corresponding value of work.)

The following is a typical extract from the preliminaries section of a bill of quantities where a fluctuating price based on the formula is required.

Formula Rules	40.1.1.1
Rule 3	Base month April 1985.
Rule 3	Non-adjustable element 10%.
Rules 10 and 30(i)	Part 1 of Section 2 of Formula rules to apply.
Clause 40 Adjustment of Contract sum – NEDO Price Adjustment Formula of Building Contracts.	Attention is drawn to the Standard Form of Building Contract Formula Rules dated 3 March 1975 and Practice Note 18 Adjustment of the Contract Sum by means of Formulae both published by the Joint Contracts Tribunal.

The following items will be dealt with as stated below:

Rule 3	Balance of Adjustable Work as defined shall include measured work undertaken by the contractor which has not been allocated to a Work Category.
Rule 7	Structural Steelwork – adjustment shall be in accordance with alternative (i) by use of the formula set out in Part 1, rule 9 using Work Category 23.
Rule 8	'Fix Only' work – Nominated Supplies and Direct Supplies from the Employer. Adjustment shall be in accordance with alternative (ii) by inclusion in the Balance of Adjustable Work.
Rule 11a	Allocation to Work Categories will be in accordance with alternative (ii), i.e. annotation of items in the Bills of Quantities.

The following symbols and abbreviations are used:

-1 to 49 -	Work Category
B.A.W.	Balance of Adjustable Work
C.E.F.	Catering Equipment Installations Formula

It is vitally important that the estimator fully understands the foregoing, and appreciates the effect such information has upon his tender.

The implications in this particular case are as follows:

Rule 3 'Base month April 1985'

The 'base month' is normally the month preceding that during which tenders have to be returned. Thus, the index numbers published for each work category for April 1985 reflect the rates of wages and prices of materials, plant, transport, etc. which are current at that particular time and which the estimator has used to price the bill.

As the work proceeds on site, the index numbers for work completed in each work category applicable at the time of each valuation are compared with those applicable at the 'base month'. The difference in the two sets of index numbers subsequently forming the basis of the calculation for the recovery of fluctuations.

Rule 3 'Non-adjustable element 10%'

Referring back to Clause 39, it has been shown that the operation of this clause involves the contractor in a 'shortfall' or under-recovering of increased costs (the estimator, however, still being able to make an allowance for this within his tender).

Again, even with the use of the formula method, the contractor could still find himself being involved in a similar sort of 'shortfall'.

It is generally agreed that the formula method is fairer and more equitable to the contractor by reimbursing him with all increases, including an element of overheads and profit. (Since the indices apply to all the main contractor's prices and since profit and overheads are an integral part of those prices, the adjustment of the value of work completed at each valuation will automatically give the contractor a proportionate increase in these two categories.) However, local authorities and central government have deemed it necessary to reduce in some way the amount of the increases payable. The amount by which increases paid to the contractor falls short of the amount originally calculated by the use of the indices is termed **'the non-adjustable element'**. This non-adjustable element is expressed as a percentage and is currently 10 per cent (as can be seen in this example), having previously been as much as 15 per cent.

In simple terms, this means that the contractor is able to recover only 90 per cent of his calculated increases.

It is reasonable to expect that the contractor, as in the case of clause 39, will want to include a compensating amount within his tender.

Thus, the estimator must make a calculation, again similar to the one made under clause 39 except that in the case of the formula he does not need to make any adjustment for the non-recovery of profit, overheads, plant, supervision, transport, etc. A typical example of such a calculation would be as follows (using the anticipated increase from the calculation of the shortfall under 39, pages 157–6. except that in this case the figures are assumed to include overheads and profit):

Anticipated increased costs will be:

(a)	Materials	£12,615
(b)	Labour	7,912
(c)	Preliminary items	3,960
		£24,487

The expected 'shortfall' due to the non-adjustable element of 10 per cent applying

= £24,487 × 10% = £2,449

By adding the £2,449 to his tender, the contractor will not recover exactly the amount of his shortfall, since the formula will also apply to this figure once it has been incorporated into his tender, i.e.

Anticipated increases	£24,487
Add adjustment for 'shortfall'	2,449
	£26,936

Due to the non-adjustable element being 10%,
contractor recovers £26,936 × 90%
= £24,242

Thus, by adding a straight 10 per cent, an under-recovery of £24,487 – £24,242 = £245 is still obtained.

If the estimator wishes to eliminate this, he should add a percentage in the region of 11 per cent, i.e.
£24,487 × 11% = £2,694.

Adjustment is now as follows:

Anticipated increases	£24,487
Add adjustment	2,694
	£27,181

The contractor recovers 90% of this, as before

£27,181 × 90% = £24,463

This, compared with the amount of £24,487 which the contractor would want to recover, all but eliminates the shortfall.

(It will not be worth the estimator's time to be involved in any more detailed calculations since the whole exercise is based on arbitrary assumptions in the first instance.)

Part 1 of Section 2 of formula rules to apply

This deals with the allocation of work contained within the bill of quantities to 'work categories'. Under Part 1 of the formula rules, such work is allocated to 34 relevant categories in series 1 and 49 categories in series 2.

(Part 11 of the formula rules uses the 'work group method' of allocating items in the bill. Using this method, work categories are combined to produce work groups, e.g. Work Group A – work categories 1, 3 and 6.) This information is more related to the valuation of the work at a later date and should not affect the pricing of the bill at tender stage.

Balance of adjustable work

There are items within the contract which do not fall into a work category. Such items include: Preliminaries, overheads and profit (if shown separately), water, insurances, bond (if any), attendances, etc. However, increased costs are still allowed on these items by applying a percentage increase to the value of them which is equivalent to the average increase of all the work categories combined.

Rule 7 – Structural steelwork

Adjustment of the value of structural steelwork within the bill may be by allocation to a work category (23) in the normal way or by the use of the specialist installation formula.

Usually, the latter method is adopted only when there is a significant amount of framed steelwork on the job, since the specialist formula is much more comprehensive than the work category method.

In this case, however, it can be seen that the allocation to work category 23 method is to be used, the amount of steelwork in the bill being relatively small.

Rule 8

This deals with 'fix only' work. In this case it is to be included in the balance of adjustable work, which is self-explanatory.

Rule 11a – Allocation to work categories

Work within the bill can be allocated to work categories (i) by the inclusion in the preliminaries of a schedule listing the 34 (or 49) work categories and the references of all bill items which will fall in each category, or (ii) as in this particular case by annotation of items in the bill of quantities. For example:

Hardcore or the like - 2 -
Hardcore
Average 150 mm thick; depositing 810 m^2
and compacting in layers

The example indicates that the hardcore bed falls in work category 2. 'Hardcore and imported filling'.

Work category

A work category is a term used for a section of work. These sections are based on the trades within the *Standard Method of Measurement*. Most of the items which comprise the bill of quantities fall into one or other of the work categories applicable, the remaining items being dealt with under 'balance of adjustable work' or by the use of a specialist formula.

(Series 1 contains 34 work categories while Series 11, being much more comprehensive, contains 49 – see Appendix A to the 'Formula Rules'.)

The pricing of the bill of quantities with the formula in operation presents a simpler task for the estimator compared with the conventional method since much of the risk of allowing for non-recoverable increases is eliminated.

The main points to bear in mind when using the formula are:

1 The contractor enjoys a far greater recovery of increased costs since he can recover increases in overheads, profit and preliminary items.
2 The contractor does not recover ***actual*** increases since the formula gives him an average recovery, the reasons being as follows:
   (i) Difference in ratios of labour : materials : plant between those in the indices and those applicable on a particular job.
   (ii) Difference between the materials used on site and those used in the compilation of the indices.
   (iii) Variations in basic and increased rates and prices used in pricing the bill and those used in the indices due to regional differences.
3 The costly administrative work in respect of invoices and labour returns associated with the conventional method is reduced drastically.
4 An allowance to offset the amount of 'shortfall' is still required, even under the formula, since on local authority contracts a 'non-adjustable element' will apply.

(* *Note*: The Series 1 work category method ceased to function in December 1981 – all 'formula' work is now carried out in accordance with the Series 2 schedule.)

Chapter 6

Preliminaries

The items found in the Preliminaries section of a bill of quantities are usually the most difficult and arbitrary of all to price. Indeed, if all the tenderers' priced bills for any one project could be examined, this section would produce the greatest variation in prices with each estimator having his own idea as to the scale and extent of the costs involved. The nature and the conditions of contract of every job are different and the items encountered in the Preliminaries are therefore necessarily unique to each particular project, and whilst records from previously priced jobs will be useful and will assist in some measure in estimating the cost of the current set of Preliminaries, they can, at best serve only as a guide and each Preliminary item must be looked at individually and priced on its own merits. There is often a great temptation, particularly when time is short, to look back at a previously priced set of Preliminaries which appears on the face of it similar to that under consideration and extract prices for use in the current job. However, whereas one set of Preliminaries does indeed often look pretty much like any other, careful examination will no doubt reveal extra restrictions and limitations imposed upon the contractor which, if overlooked, could lose him a great deal of money. Since each set of Preliminaries is different from any other, the best way of illustrating the pricing of this section of the bill is to consider a typical set of Preliminaries (Example 1) for a specific contract and this chapter is therefore devoted to that end.

The project in question relates to the proposed construction of a branch library for a local authority executed under the JCT Standard Form of Contract Local Authorities (with quantities) 1980 edition. Those clauses which have an effect on cost at tender stage are discussed in detail in chapter 4.

Those documents relating to the contract and from which the total cost of the preliminaries will be determined, are reproduced on pages 186 to 239.
These are:

Drawings showing site layout, plans and elevations.
The Preliminaries section of the bills of quantities.
A contractor's typical standard form for pricing Preliminaries prepared by himself to suit his own particular circumstances.

6.1 The drawings

The drawings (plans and elevations at least) would form part of the tender documents and would serve to give the estimator an overall picture of the scale and complexity of the job which often is not apparent from the bill of quantities alone. The drawings themselves would have a major influence on the pricing of the work.

From the drawing for the library, it can be noted that:

1 The site is in the middle of an existing busy shopping precinct and therefore great care must be taken to protect passers-by from site activities and the site itself must also be made secure from damage and vandalism. Being situated in a busy shopping centre in an inner city area, the site could be vulnerable – probably an expensive item.
2 The site is approached directly from the main service road to the shopping centre and a crossover must therefore be formed and the road and footpath kept free from mud, etc. during the course of the works.
3 Existing services, drains, electricity and water are conveniently situated nearby as shown, although the connection into the sewer in the road will cause traffic congestion and some form of control will be needed.
4 The area of the site surrounding the building itself appears to be adequate for all storage needs, stockpiles of materials, top soil and site huts, etc.
5 The existing trees will form part of the final landscaped scheme and are to be retained. In addition to the cost of erecting some form of fencing to protect them, the storage area for materials is reduced further.

6 Access to all parts of the building seems to be good. This is an important point since the roof is a steel-framed structure and the first floor is constructed from precast concrete beams which calls for the use of a crane. Hardstanding areas of hardcore for the crane can be formed quite easily thus enabling hoisting operations to be carried out from any position

6.2 The Preliminaries section of the bill

The Preliminaries are items which set out to describe the requirements peculiar to the job in question, the nature of the site, description of the works, names of the parties involved, conditions of contract and so on. The items to be included in the Preliminaries section of the bill are listed in Section B of the Standard Method of Measurement. As has previously been stated, this part of the bill is the most difficult to price, but in this example each item will be looked at and priced in detail and can be regarded as the only accurate method of establishing the cost involved, although some contractors still resort to allowing a lump sum for all the preliminary items based on a percentage of the contract value. Whilst an idea of the *approximate* cost can be obtained in this way by using previously priced similar jobs as a yardstick, it remains, at best, an extremely rough and ready method and is not recommended!

6.3 Contractors' standard schedule for pricing

Illustrated on pages 187 to 209 is a *typical* contractor's schedule for pricing preliminaries. Included are those items which will occur on most contracts, although space is provided for extra items to be written in when required. It is also likely that not all of the items would be needed for every contract, in which case those would obviously remain unpriced. The schedule would be priced in accordance with the requirements stated in the Preliminaries and a total lump sum figure obtained which is then transferred to the tender summary.

Individual prices need to be allocated to the appropriate items in the Preliminaries bill only if the contractor's priced bill of quantities is called for.

6.3.1 Method of pricing preliminary items

The following section deals with a description of the preliminary items to be found on pages 176 to 185 together with a brief explanation of how the price may be determined, with the detailed calculations being shown in the contractor's own pricing schedule referred to earlier.

(a) Descriptive items

Not all of the items in the Preliminaries bill will need pricing since many of them are purely descriptive and are intended to give the contractor an overall picture of the work. Examples of such items are as follows:

1 *The Preliminary particulars* – names of the parties, description of the site, description of the works, submission of tenders.
2 *Definitions and qualifications* – definitions, quantities, cost analysis information, pricing of Preliminaries, rates in the bills of quantities, all these items being contained in pages 1/1–1/5(B).

(b) Conditions of contract – pages 1/4(D), 1/5(c) and 1/6–1/9

The form of contract is always clearly stated together with a brief schedule of the clause headings which are to apply, including those clauses which are to be amended or deleted for this particular job and the details which are to be inserted in the appendix to the conditions.

With regard to the conditions of contract, most clauses will have no effect on cost at this stage and so will not be priced; however, those which will need pricing are discussed in detail in chapter 5.

(c) Other obligations and restrictions imposed by the employer

In addition to the obligations imposed by the conditions of contract, there may be further obligations and/or restrictions imposed by the employer and it is likely that such items will involve the contractor in further expense. Examples of such items are as follows.

1. The provision of a guarantee bond – page 1/10 (ii). This is not required by all employers since the cost of providing the bond will be incorporated into the tender figure which means that the employer himself is paying for it, but it is a requirement favoured by local authorities and other public bodies. In this case the contractor is asked to take out an insurance with his insurance company or bank for an amount equivalent to 10 per cent of the contract sum (assuming the contract exceeds £100,000 in value). This ensures that should the contractor be unable to fulfil his contractual obligations, the guarantor will either:

(a) Pay damages to the employer up to the full value of the bond (i.e. 10 per cent of the contract sum); or
(b) Arrange for another contractor to finish off the outstanding work, again, up to the value of the bond. Where the cost of completing the work exceeds this figure, the employer himself must bear the additional expense. He could, of course, request the contractor to provide a bond to the full value of the contract, but the cost of the premium would probably be so high as to inflate the tender beyond what the employer is prepared to pay.

The amount of the premium will vary depending on the risks involved in the contract and on the financial standing of the contractor himself and a quotation would be obtained by the estimator (probably through the company secretary) and included against this item. Premiums are in the region of about 6 per cent of the value insured, or in this case 0.6 per cent of the contract sum; parts (i), (iii), (iv) and (v) of this item on pages 1/7 and 1/8 would be for information only and do not require pricing.

2. *Provision of temporary accommodation – page 1/25 item A*. A site office is usually required by the employer for the use of the architect and clerk of works. In this case the employer has defined exactly what his requirements are to the extent of including the provision of office furniture, heating, lighting and washing facilities. (Whether in fact the accommodation which is actually provided meets these specifications remains to be seen!).

The method of pricing this item would be as follows:

(a) Price for loading the site hut (whether ready-assembled or sectional type) on to a lorry at the contractor's yard.
(b) Allow for haulage cost to and from the site.
(c) Allow for unloading cost on arrival at the site.
(d) Allow for erection (depending on the type of accommodation, allow, say, 3 joiners and 2 labourers for half a day plus sundry materials and paint for repairs).
(e) Hire cost per week × number of weeks of contract.
(f) Allow for the cost of dismantling on completion of contract, haulage back to the yard and all handling costs as before.
(g) Cost of providing office furniture as described.
(h) Hire of Calor gas heater (or similar). Hire cost per week × number of weeks of assumed use (remembering to exclude the summer period!).
(i) Say, 1 gas bottle per week of use.
(j) Cost of attendance, sweeping out, changing gas bottles, etc. – say, 1 hour per week for a labourer.

(The cost of the electricity and water supply can be more conveniently priced with the contractor's temporary works later on.)

3. *Provision of temporary telephone – page 1/25 item B*. Where the employer wishes to make use of the contractor's temporary site telephone, the cost should be covered by a provisional sum (see SMM6, B.8K). However, sometimes, as in this case, the wording of the 'Temporary Telephones' item puts this burden on to the contractor and unless read carefully this extra cost could easily be overlooked. The employer does not require an exclusive line here, but wants an extension to the clerk of works office with the contractor paying all 'charges and expenses in connection therewith'

(i.e. *including* the cost of all calls made!). Again this can be taken into account when estimating the cost of the telephone service for the contractor's own use later on (see page 197).

4. Safeguarding the works and protection of adjoining properties, trees, etc. – page 120. Considering the location of the contract, it is likely that the contractor will choose to erect a secure fence around the whole perimeter of the site with sturdy lockable gates at the entrance and possibly a fenced-off compound for dumper, plant, saw-bench, etc. within the site itself.

The method of pricing is as follows.

(a) Calculate the cost of materials, posts, chain-link fencing, hoardings, etc. (Full cost may be shared between several contracts if the materials can be salvaged for re-use.)
(b) Allow for fabrication costs together with the cost of handling, transporting to site, unloading and fixing in position, including any re-siting during the course of the contract if necessary.
(c) On completion, allow for taking down, making good to damaged pavements, etc. and transporting materials back to the yard, remembering to take into account the possible increased cost per hour of the labour and plant involved where the contract is a firm price for a long period of time.

Other obligations imposed by the employer, but involving no cost would be: page 1/22 Item A; page 1/22 Item B; page 1/23 Item A; page 1/23 Item B; page 1/24 Item B.

Further obligations which could be imposed but not included in this contract but well worth a mention here could include:

1 *Limitations of working space*. This may involve uneconomic working due to shortage of storage space which may hamper site efficiency and involve higher transport charges and double handling where some materials may have to be first delivered to the contractor's yard and stored there for a while before they can be delivered to the site.
2 *Limitations of working hours*. The contract may involve extensions to a bank or hospital say, where perhaps work must be suspended for part of the day to allow operations or other activities to proceed unhindered by noise, etc. Clear details would have to be given of the restrictions involved which could necessitate the contractor working only a 6-hour day whilst having to pay his men for 8 hours. The cost of this would either be incorporated into the all-in rate for labour or could be calculated and priced as a lump sum under this heading.
3 *Access*. May be restricted in some way, meaning certain items of plant may not be able to gain access to the site thus involving the

use of alternative, and possibly less efficient, methods. The extra over cost, if any, could be built up as a lump sum or included within the affected bill rates themselves.

(d) Provision of general facilities by the contractor and other obligations

Items which appear under this heading will normally need pricing by the estimator and for his convenience are usually grouped together, this section producing the major part of the cost of preliminaries. The method of pricing these items is as follows.

1. Plant, tools and vehicles – page 1/19 item A. This will be an expensive item and accurate pricing demands consultation with the plant manager and contracts manager in determining the method of working and the type of plant needed. This information and the length of time required on site for each item of plant can be extracted from method statements and the preliminary bar chart type programme.

(a) Establish the hourly or weekly cost of the item of plant being priced including the operator where appropriate.
(b) Allow for the cost of handling, haulage to site and setting up.
(c) Calculate hire rate × number of days or weeks anticipated on site.
(d) Allow for attendance (if any) where not allowed for in the unit rates.
(e) Allow for the cost of dismantling, handling and haulage back to the yard after use.

Where an item of plant is required intermittently, it must be decided whether it is cheaper to have the machine 'off-hire', involving the cost of transporting to and from the yard again at a later date or to leave the plant idle on site, incurring hire charges until it is needed again. It is assumed that the contractor wishes to price his 'plant' items here as a lump sum in the preliminaries. Alternatively, he has the choice of either pricing the appropriate plant items at the commencement of each trade within the bill of quantities or including the cost within the unit rates. The problem with these methods, however, is that accurate allocation of cost becomes impossible where an item of plant or equipment is shared between a number of trades, e.g. scaffolding would be used by bricklayers, roofers, plasterers, etc., as would a hoist, mixer and so on. However, this is a matter of opinion and each estimator will have his own preferred method of pricing

2. Safety, health and welfare of workpeople – page 1/19 item B. The cost of protective clothing, goggles, boots and gloves, etc.

can be calculated as a lump sum, with the provision of canteen facilities, drying room, washing facilities and first-aid equipment priced under 'temporary accommodation' later on.

3. *Site management costs – page 1/19 Item C.* All off-site management costs are assumed to be included in the head office (overhead) charges (see chapter 3). On-site management and supervision costs are allowed for under Clause 10, 'Person in charge' (see chapter 5, page 136).

4. *Labour on costs – page 1/19 item D.* All items under this heading, with the exception of travelling expenses, can be taken into account when building up the all-in rate for labour.

It is usually more convenient, however, to allow for the cost of transport and travelling for operatives at this point. The method of building up the cost of travelling will depend on the contractor's own policy with some providing transport for the men from, say, Head Office to the site, and others preferring to pay fare allowances with operatives finding their own way to the site. Under Rule 14 of the NWRA, operatives are entitled to two payments when no transport is provided by the contractor, namely: a 'fare allowance' to cover the cost of transport and a 'travel allowance' designed to reimburse the man for travelling in his own time. The allowances, based on a sliding scale and dependent on the distance of the operative's home from the site are stated in the table forming part of the agreement and are fixed sums of money per day irrespective of the mode of transport used by the operative.

Thus in the case of a contractor providing no transport, the estimator must calculate the average weekly site labour strength, obtained from the total estimated man hours in the contract, and then decide on how far, on average, each man will have to travel. This is termed the 'importation of labour' and is necessarily only an approximate estimate since it is impossible to decide at this early stage exactly who will be assigned to the site and for how long. The estimator may therefore assume that, say, 40 per cent of the men live locally (i.e. less that 6 km from the site) and therefore no allowance is paid, 20 per cent live 10 km from the site, a further 20 per cent live 20 km away, 10 per cent live 30 km and 10 per cent live more than 50 km away and therefore qualify for lodging allowance. Having made this assumption, the total cost of travelling can then be calculated using the latest figures in the published tables.

5. *Maintenance of roads etc – page 1/21 item A.* Allow, say, ½ hour per day for a labourer keeping the road and public footpath in the vicinity of the site clean and free from mud and debris often deposited by lorries entering and leaving the site. It would also be prudent to include a sum for reinstating an area of the tarmac

footpath at the site entrance and for the replacement of a number of road kerbs which inevitably will be damaged during the contract. Photographs could be taken before commencement of the works where broken kerbs or pavings already exist thereby avoiding disputes later on as to whose responsibility it is to make good the damage.

6. Police regulations – page 1/21 item B – and protection of waterways – page 1/21 item C. Would not require pricing on this occasion.

7. Water for the works – page 1/23 item C. Note that the bill refers the contractor to the 'final summary', indicating that it is more suitable for pricing at the very end of the estimate after every other item has been priced rather than here in the Preliminaries section. The reason for this is that a large part of the actual cost will be dependent on the contract sum which will not be known at this stage.

Allow: (a) for the cost of the water; and (b) for the cost of the labour and materials involved installing the temporary supply and clearing away on completion.

(a) *Cost of water.* The charge from the water authority will not be in accordance with the quantity used, but in relation to the contract value (except where an extensive and lengthy contract is involved in which case the supply may be metered and the contractor charged for the actual amount used). Most authorities, however, charge so many pence per £1,000 contract value, a sum which can be calculated by the estimator and included as a lump sum. Although the problem does not exist on this particular contract, sometimes the authority's water main does not extend as far as the site boundary in which case a provisional sum should be included in the bill to cover for this work.

(b) *Cost of installing temporary supply.* Fortunately, on this contract the existing water main is situated near to the site boundary and a supply can be obtained without having to dig up the road, although a trench will have to be dug across the footpath which in turn will have to be reinstated at the end of the job. Allow for builders' work excavating trenches to provide a supply to site huts, toilets and say two standpipes suitably positioned on site and for making good on completion where necessary. Include in addition a charge by the water authority for tapping into the mains and, say, a week for a plumber installing and subsequently removing standpipes, supplies to washrooms, toilets, etc., together with the cost of tubing, fittings and other sundry materials.

8. Lighting and power – page 1/24 item A. The provision of an electricity supply is vital and will be one of the first tasks the contractor will want to organise on taking possession of the site. However, it will be some time before a permanent supply can be provided (probably by the nominated sub-contractor responsible for the electrical installation) and therefore an immediate supply must be obtained from a generator.

(a) Allow the cost of haulage to and from site together with handling costs, in addition to the hire rate of the generator × number of weeks expected on site.
(b) Allow for an electrician for, say, one week installing the temporary supply for power tools, site huts, site lighting (if necessary), etc.. plus a lump sum to cover the cost of leads, bulbs, lampholders, etc., allowing again for clearing away on completion.
(c) The cost of the electricity used when the permanent supply is installed will be charged by the electricity board on a quarterly basis. Records from previous jobs can be examined as a guide to the number of units consumed on similar projects. Allow average cost per quarter × number of quarters in the contract period.

    Although not applicable here, in some cases, where refurbishing work or extensions to an existing building are concerned, the contractor may be able to make use of the employer's existing supply. If this is the case, it should be clearly stated in the Preliminaries and whether the electricity will be supplied free of charge or the cost shared in some way between the employer and contractor.

9. Temporary works generally – page 1/24 item B. This clause draws the estimator's attention to the fact that all temporary works must be cleared away on completion and damaged areas made good, and an allowance for this is made when pricing all such items. The requirement for reinstatement of top soil will depend on the nature of the site, but not wishing to 'load up' the tender at this stage it is assumed that the estimator is confident that 'reasonable care' will be taken and any cost here can be avoided (whether his confidence is justified remains to be seen!).

10. Temporary roads – page 1/24 item C. A temporary roadway of sleepers is called for under this clause; however, these are very expensive and difficult to obtain, so it is more than likely that they will only be used in forming temporary crossovers to protect existing kerbs and footpaths from heavy lorries with all other hardstanding areas and roadways being of some suitable hardcore. When pricing for hardcore roads, establish on the plan where site offices and storage areas are to be situated and calculate the quantity of material required for the area in question for a depth of, say, 200 mm. Allow for topping up soft spots, etc. during the course of the contract and for a digger and gang of three or four men laying the road and for the same plus a lorry for taking up and disposing of the material on completion.

11. Temporary buildings – page 1/25 item A. The provision of a site office for the architect has been covered earlier and the same principle of pricing would be used to determine the cost of the contractor's own accommodation. In addition to his site office, a canteen, drying room, storage huts and temporary toilets will have to be supplied. (Storage huts provided by the contractor for the use of a nominated sub-contractor would be priced for under the appropriate 'special attendance' item following the PC sum allowed for the specialist's work.) An additional cost over and above those already mentioned in the pricing of this item would be for the connection of the drains from the toilets to the main drainage system and for dismantling on completion of the works.

Rates would not be applicable for such a short-lived contract as this, but long-running contracts of two years or more duration may attract some form of charge. A telephone call to the local rating office would verify this point.

12. Temporary telephones – page 1/25 item B. Allow for the cost of installation and extension to the clerk of works' office and for a coin box and external bell (if required), plus the quarterly rental and likely cost of calls × the number of quarters in the contract period (telephone charges on a similar contract currently under construction will serve as a good guide to extent of costs involved).

13. General scaffolding – page 1/26 item A. Internal and external scaffolding will be required, with traditional putlog scaffolding being used externally and trestles and boarding internally for use by ceiling fixers, plasterers, painters, etc., together with a mobile tower scaffold for a short while to carry out work to the area of high-level ceilings. The contractor can either supply the scaffolding himself or employ a specialist firm of scaffolders, popular nowadays, to supply, erect and dismantle the external scaffolding as required. Although he will expect to pay more for this service, it does avoid the cost of maintaining large stocks of tubing, fittings and boards at his yard which if not cleaned, oiled and carefully stored will quickly deteriorate. Set against this advantage, however, is the inconvenience of not having stocks on hand readily available if needed in a hurry.

Where the contractor chooses to sub-let the scaffolding work, a minimum hire period of approximately four weeks will probably apply with a reduced rate per week applicable after the expiry of this initial hire period, the exact terms being stated on the quotation.

When the contractor's own scaffolding is to be used, the estimator will work on a rate per square metre of external face of the building using prices from previous jobs. However, initially a rate would have had to have been built up from first principles based on a typical area of scaffolding common to all work. The method of pricing is as follows.

(a) Draw a sketch of the scaffolding requirements to a typical length of wall.
(b) Calculate total length of tubing, fittings and boards.
(c) Allow a gang of three men, say, one day for erecting approximately 400 m of tubing, including boards and fittings, and about half this time for dismantling.
(d) Build up the cost of materials using current prices.
(e) Allow *about* thirty uses (this covers breakages, losses, etc.).
(f) Allow for transport costs both ways.
(g) Allow approximately four hours per week for maintenance.
(h) Extract from the programme the number of weeks the scaffolding will be required on site in order to calculate the total cost.

A rate per square metre will then be established which can then be updated or adapted for more complex jobs when pricing future contracts.

14. Works by nominated sub-contractors – page 1/26 item B. This item is purely descriptive and will not need pricing. It reminds the contractor of his obligations with regard to work carried out by nominated sub-contractors, in particular to the importance of close liaison and co-ordination of the work and the procedure in connection with payments. Any profit, general and special attendance items on nominated sub-contractors' work will be priced under the heading of 'PC and Provisional Sums'.

15. Nominated suppliers and 'fix only' items – page 1/27 item A, 1/28A. Again these items will remain unpriced, being merely descriptive clauses.

16. Protecting from the weather – page 1/28 item C. Providing protection for materials on site will have been allowed for in the pricing of temporary storage huts and compounds earlier on, whereas this item is intended to cover the works themselves where typical operations may include the provision of tarpaulins to protect brickwork from extreme weather conditions or polythene sheeting to protect concrete whilst curing, etc.

17. Drying the works – page 1/28 item C. In this case a provisional sum has been included in the bill of quantities which relieves the estimator from the risk of pricing an item which could be either extremely costly or negligible depending on the type of construction and the time of year the job is scheduled for completion. In this case, therefore, the item will remain unpriced, but SMM6 requires it to be priced in the tender by the contractor (Clause B.13 1r) which means the estimator must decide on what form of heating can be used and for how long it will be needed. Sometimes the building's own heating system can be utilised where it has already been commissioned and working at, say, a low test rate, in which case the correct slow rate of drying out can be achieved, the contractor no doubt being only too pleased to contribute towards the cost of the fuel used. However, where this is not possible, the estimator must allow for Calor gas burners or something similar together with a supply of gas and maybe half an hour per day for an operative in attendance checking the equipment and changing the gas bottles. Where the employer requires a specific temperature and humidity to be maintained for some special purpose such as the laying of an epoxy resin floor finish, say, a separate item should be stated and the contractor given the opportunity of pricing for this extra obligation (SMM6, Clause B.8 1h)

18. Removing rubbish and cleaning – page 1/28 item D. This requirement calls for the works to be kept tidy and free from rubbish, etc. during the course of the contract and for a thorough clean down prior to handover on completion. Whilst it may be expected that trades will clean up after themselves as the work proceeds, in practice this is often not the case and the estimator has to allow for a clean up and removal of rubbish at regular intervals during the contract – allow, say, a digger plus a lorry and gang of four labourers together with tipping fees (where necessary) for half a day each month.

Before the building can be handed over to the client on completion of the works, the contractor must mount a serious (and expensive) effort to clean and wash down walls, floors, windows, paintwork, etc. in order to prepare for its occupation – no easy task after thirteen or more months of building operations. In addition to the cleaning down operation, all protective casings, packaging, etc., such as adhesive tape surrounding metal windows and edges of sanitary ware, must be removed and disposed of as well as all the accumulated little piles of debris and rubbish scattered over the site and previously overlooked! – allow a digger and lorry as before for say two days and a gang of four men for one week plus sundry cleaning materials, brushes, etc.

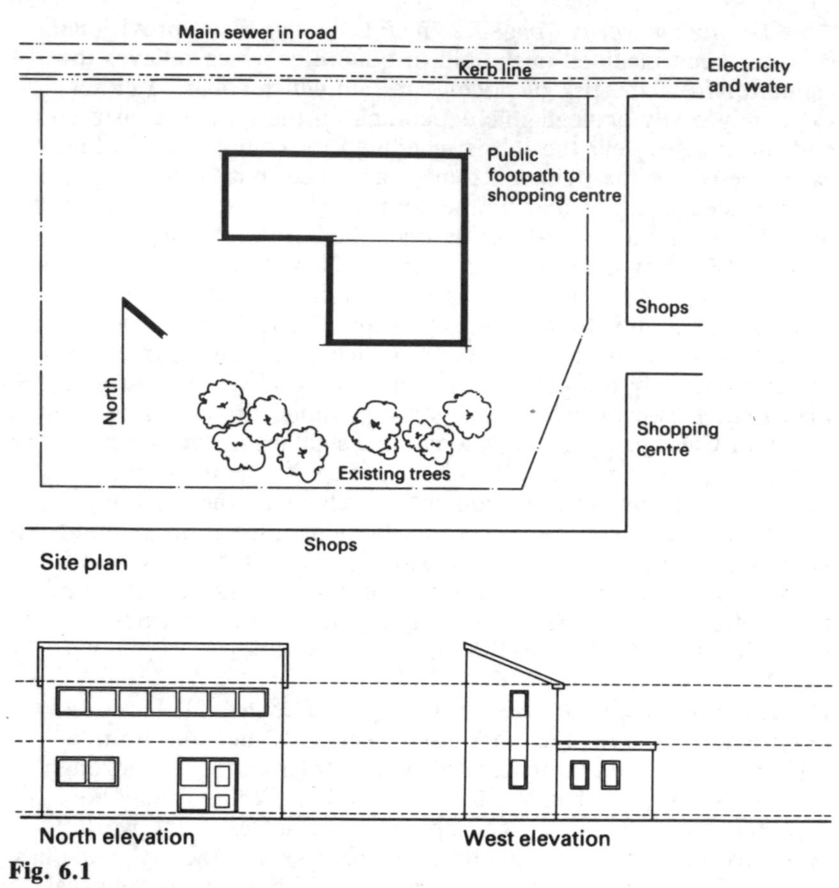

Fig. 6.1

The following reproduction of the contractor's own pricing schedule for preliminaries uses the methods discussed here in order to determine an actual figure for this section of the bill. However, it is stressed that this particular procedure is only *typical* of the methods used and each estimator will develop his own individual system designed to suit his particular circumstances.

TENDER FOR *construction of Branch library*

At Kingsway shopping precinct, Southtown

ARCHITECT	*A.B. James R.I.B.A.*
QUANTITY SURVEYOR	*J.D. Hall & partners ARICS*
CLIENT	*Southshire County Council.*
DATE FOR SUBMISSION	

Build up of transport

Item A

Offices required: 1 site off., 1 c/w office; 2 stores; 1 canteen; 2 toilets; 1 compound. Lorry can transport 2 huts per journey plus 1 journey for compound alone = 5 round trips delivering huts × 2 for bringing back to yard on completion
= 10 round trips

Allow: 1 hour loading at yard
1 hour unloading at site
2 hours journey to site
2 hours back to yard

	6 hours × £11 (lorry and driver)	= £66
allow:	2 labourers at yard loading	
	1 hour × £4 × 2	= £8
	2 labourers at site unloading	= £8
	Therefore cost per round trip	= £82 × 10 trips = £820

Items B and C

Items needed:

Dumper	– 1 trip
2 mixers	– 1 trip
Hoist	– 1 trip
Compressor	– 1 trip
Saw bench	– 1 trip
Hard roller	– 1 trip
Generator	– 1 trip
2 pumps	– 1 trip
2 vibrators	– 1 trip

9 round trips × 2 for bringing back to yard on completion.
= 18 round trips × £82 = £1,476
(Mobile crane hired locally – no transport charge here)

Item F

Allow 2 trips per month for sundry materials delivered to site.
13 month contract × 2 trips = 26 trips × £82 = £2,132

Transport

		£
A	Transport temporary accommodation	820
B	Transport mechanical plant	
C	Transport small plant	1,476
D	Transport scaffolding (*see scaffolding build up*)	--
E	Site lorries clearing spoil (*incl. in excavation rates*)	--
F	Site lorries delivering material	2,132
G	Tractor and Trailer (*N/A*)	--
H	Dumpers (*see mech plant section*)	
I	Buses coaches – *no transport provided – fares and travel allowances paid*	--
J	Vans	--
	C/F to summary	£4,428

Build up of plant and hoisting

	£
Item A (from programme)	
Allow on average 2no. 5/3½ mixers for 35 weeks = 70 weeks at hire rate £30 per week all-in	= 2,100
Item B	
Telescopic mobile crane for steel roof beams and precast concrete floor – 1 week for steelwork + 1 week for floor = 2 weeks at £20 per hour (incl. driver) = 80 hours £20 (cost of fixing gangs incl. in unit rates in BQ)	= £1,600
Item C	
Electric hoist req'd from 1st lift of brickwork to completion of roof coverings (week 17 to week 40) = 23 weeks at £40 per week	= £920
Item D	
One 10 KVA generator for 10 weeks until electricity supply installed = 10 weeks at £40 per week	= £400
Item E	
4-tool compressor for 1 month (b/out road and footpath for services etc.) = 4 weeks at £50 per week	= £200
Item F	
Excavations carried out in winter. ∴ allow 2no. 100 mm. dia. sludge pumps and hoses to keep excavations clear of surface water for total excavation time (from programme) = 2no. for 10 weeks at £18 per week	= £360
Plus ½ day maintenance = 2½ hrs per week × 10 weeks at £4 per hr	= £100
Allow for fuel (say)	= £50
	£510
Item G	
1no. from week 25 to week 40. = 15 weeks at £27 per week	= £405

Mechanical plant and hoisting

		£
A	Mixing plant	2,100
B	Cranes	1,600
C	Hoisting and lifting equipment	920
D	Generators	400
E	Compressors and tools	200
F	Pumps and hose	510
G	Portable saw bench	405
H	Kango hammers (*incl. with compressor*)	—
I	Concrete vibrators (*incl. in unit rates for conc.*)	– –
J	Cement silos (*N/A – Ready mixed conc. used*)	– –
K	Roller (*incl. in unit rates for excav. work*)	– –
L	Dumpers, *1no. 15 cwt dumper (on ave.) for 30 weeks at £35 per week (all-in)*	1,050
M	Disc cutter (*for chases, holes etc*) *2no. for 1 month at £15 per week = 8 weeks at £15.*	120
	C/F to summary	£7,305

Build-up of labour, materials, installing and clearing plant

	hrs
Item A	
2 labourers × 2 hrs each mixer	= 8
Item B	
N/A mobile crane used	—
Item C	
4 labourers × 4 hours	= 16
Item D	
2 labourers × 2 hours	= 4
	28 hrs
	28 hrs × £4 = £112

Labour and mats, installing and clearing plant		£
A Mixing plant		
B Cranes (N/A)		
C Hoists and lifting equipment		112
D Portable saw bench		
E Cement silos (N/A)		--
	C/F to summary	£112

Build-up of site accommodation

	£
Item A	
56 weeks × £20 week hire rate (haulage and handling incl. in transport)	= £1,120
Item B	
56 weeks × £23 per week hire rate	= £1,288
Item C	
56 week × 2 store huts = 112 weeks × £15 per week	= £1,680

Item D

Mats to form compound – area 20 m × 10 m
100 × 75 posts $2\frac{1}{2}$ m long – 32no. at £160/m³ = £96
1.8 m high chain link fencing – 10 no. 10 m rolls at £30/roll = £300
Timber for gates, ironmongery etc. (say) £50
Total mat. cost = £446 – assume 3 uses

$\therefore$ cost per use = $\frac{£446}{3}$ = £149

Labour:
Gang of 2 joiners + 1 labourer 4 days erecting = £4.50 + £4.50 + £4 = £13/hr × 32 hrs = £416
Gang of 3 labourers 1 day dismantling = 3 × £4 = £12/hr × 8 hr = £96
total lab. cost = £512

Total cost = £661

Item E

Hire of 2no. portable toilets at £10 per week for 56 weeks = £1,120 + pipework connecting to main drain (say) £50 plus 2 labourers for 1 day making connection and dismantling on completion = 2 × £4 × 8 hrs = £64

Total cost = £1,234

Item F

56 weeks × £18 per week hire rate = £1,008

Item G

Gang of 4 labourers + 2 joiners for 1 week (including erection and striking) = (4 × £4) + (2 × £4.50) = £25/hr × 40 hrs = £1,000 plus sundry materials, paint etc. for repairs (say) £100 – total cost £1,100

Item H

Labourer ½ hr per day = 2½ hrs/weeks at £4 hr = £10.00 per week × 52 weeks = £520

Item I

4no. calor gas heaters for 8 months £20 per month ea. = £640

(includes supply of gas bottles)

Item J

10 prs. boots at £5
4 prs. goggles at £4
15 helmets at £6
10 sets waterproofs at £9
= £246

Site accommodation

			£
A	Site offices		1,120
B	Clerk of Works office		1,288
C	Stores		1,680
D	Compounds		661
E	Sanitary accommodation		1,234
F	Canteens		1,008
G	Erection and striking		1,100
H	Cleaning and attending		520
I	Stoves and fuel		640
J	Protective clothing		246
K	Site office equipment		
L	Clerk of Works office equipment	*incl. in hire rates*	– –
		C/F to summary	£9,497

Build-up of temporary services reinstatement and sundries

Item A £

Electrician wiring up site huts, site lighting etc. and taking down on completion 1 week at £6 hour = £240
mats: lamps, cables, leads, etc. (say) £200. Electricity charges – 5 quarters at £150 per quarter = £750
Total cost = £1,190

Item C and F

Plumber providing temp. supply to huts and standpipes plus dismantling on completion 6 days at £6 per hour = £288

Labourer digging trenches and attending – 2 days at £4 = £64
Materials – pipework, fitting, taps, etc. = £200

Water authority charge: for drilling and tapping main £150
for building water based on
Contract sum – see tender summary. Total cost = £702

Item D

Installation charge	£100
Extension to C/W office	£40
External bell	£35
Coin box	£40
Rental – 5 quarters at £50	£250
Calls (say) £100 per qtr.	£500

Total cost = £965

Item E

Sign board – 2 sheets plywood at £20, posts, bracing (say) = £40
Fabrication, joiners and signwriting (say) = £70
Erection and dismantling shop: = £120
On-site – 2 labs × 4 hrs = 8 hrs at £4 = £32
Concrete etc. for post hole = £30

Total cost = £292

Item G

Provisional sum. included – see page 1/28

Item H

Temp. fence –
Lab: gang of 4 labourers erecting and striking on completion, 2 weeks.
= 4 × 39 hrs × £4/hr = £624

Mats:

30 rolls chain link fencing at £30	= £900	
70 posts at £5 ea.	= £350	
10 rolls barbed wire at £10	= £100	
timber for gates (say) £100	= £100	
joiner making gates 1 day at £4.50/hr.	= £36	
	£1,486	

Assume 3 uses for mats
Therefore cost per use = £1,486 ÷ 3 = £495

Therefore Total cost = £624 + £495 = £1,119

Item I

0.6% of contract sum – see final summary

Temporary services reinstatement and sundries		£
A	Electric supply services & reinstatement	1,190
B	Gas supply service & reinstatement (*N/A*)	--
C	Water supply service & reinstatement	702
D	Telephone	965
E	Hoardings and sign boards	292
F	Temporary plumbing (*incl. in item C*)	--
G	Drying out the works (*Provisional sum*)	--
H	Security watching and lighting, fences	1,119
I	Contract bonds (*see summary*)	--
J	Special risk policies (*N/A*)	--
	C/F to summary	= 4,268

Build-up of defects & extra labour costs £

Item A

See chapter 5 ('Contract clauses and their effect on cost' – pages 139–42).

Item B

Average site labour force = 15 men (excl. 5/C) total no. of weeks on site = 56 less holidays = 56 weeks – (say) 6 weeks = 50

Assume 40% of men live locally (ie n.e. 6 km) NIL

20% live 10 km from site:
= 3 men × 5 days × 50 weeks = 750 man days.
Travel allowance = 24p/day, fare allowance = 36p/day
= 60p per day per man. Total cost = 60p × 750 = £450

20% live 20 km from site:
750 man days (as above) × (84p travel all. + £1.12 fare all)
= 750 × £1.96 = £1,470

10% live 30 km from site:
(say) 2 men × 5 days × 50 weeks = 500 man days
500 man days × (£1.44 travel all. + £1.66 fare all.)
= 500 × £3.10 = £1,550

10% live more than 50 km from site and therefore qualify for lodging allowance at £6.25 per night.
= 500 × £6.25 per night = £3,125

Total cost of importing labour = £6,595

(travelling and fare allowances are subject to frequent review, and care should be taken that figures from current tables are used.)

Item C

NIL

Item D

Included in 'B' above

Items E, F

Allowance made in the 'all-in' rate

Item G

(say) 20 concrete cubes for testing at £5 ea = £100
sample panels of facing brickwork (say) 150 bricks at £120 per 1.000, labour 4 hours at £6.50 + mortar (say) £5 = £49

total cost = £149

Item H

100 pegs at 450 mm long 40p each = £40
20 profile boards at £2 = £40
Nails, lines etc. (say) £20 = £20
Foreman – incl. in I below.
Labs. assisting (say) 2 labs for 3 days = 48 man hours at £4 = £192

total cost = £292

Item I

Foreman – 56 weeks at £140/wk £7,840
Company car and expenses 56 weeks at £100 = £5,600
Bricklayer foreman for $\frac{1}{2}$ contract assuming 50% of time supervising
= 28 weeks × £100/wk × 50% = £1,400
Joiner foreman – same as above = £1,400
Ganger for 20 weeks 25% of time supervising
= £90/wk × 20 weeks × 25% = £450
No site clerk or storeman.

total cost = £16,690

Defects liability and extra labour costs

		£
A	Defects liability	400
B	Travelling time	6,595
C	Expenses (*N/A*)	– –
D	Lodging allowance (*incl. in item B*)	– –
E	Abnormal overtime (*N/A*)	– –
F	Guaranteed week (*N/A*)	– –
G	Testing materials	149
H	Setting out	292
I	Foreman	16,690
	C/F to summary.	= 24,126

Build-up of cleaning site and clearing rubbish

	£

Items A and B

4 labs. ½ day per month plus machine plus lorry.

Hourly rate: 4 labs at £4/hr = £16.00
lorry and driver = £11.00
machine and driver = £12.00

£39.00/hr × 4 hours
= £156 per month

Add tipping fees (say) £40 = £196/month for 12 months
= £2,352

Items C

Incl. in A and B above.

Items D and E

Gang of 3 labourers for 1 week = 3 × 39 man hours
= 117 man hours at £4/hr. = £468
Machine and driver – 2 days at £12/ph. = £192
Lorry and driver – 2 days at £11/ph. = £176
Tipping fees; (say) £40 = £ 40
Cleaning materials and equipment (say) £100
total cost = £876

Item F

Hardcore for road and hardstanding areas.
150 m × 3 m wide × 150 mm thick = 68 m³h/core
= approx 150 tonnes at £10 per tonne = £1,500
Laying roads – JCB and driver plus 3 labourers for 2 days
Hourly rate = £12 + (3 × £4) = £24 × 16 hrs = £384
Filling potholes and soft spots (say) 1 load every 2 months
= 6 loads × 10 tonnes = 60 tonnes at £10/tonne = £600
Labourer filling potholes (say) 1 hr/tonne = 60 hrs at £4 = £240
Clearing away on completion – lorry and driver plus
JCB and driver plus gang of 3 labourers 1 day = £192
Tipping fees (say) £40
Total cost = £2,966

Item G

1 hour per week for 1 labourer
= £4/week × 50 weeks = £200

Item H

Allow 10no. railway sleepers at £10 ea. = £100 plus 2 labs for 2 days reinstating kerbs, footpaths, etc. = 32 man hrs × £4/hr = £128

Mats:

Kerbs, tarmac, concrete (say) £100

Total cost = £328

Cleaning site and clearing rubbish

		£
A	Clearing surplus material	
B	Cleaning out during construction	2,352
C	Cleaning out after trades	
D	Cleaning out at completion	976
E	Clearing rubbish	
F	Access roads incl. cleaning and maintaining	2,956
G	Keeping roads footpaths clean	200
H	Repairs to kerbs etc.	328
	C/F to summary	6,812

Build-up of scaffolding and gantries

Item A

Assume typical putlog type scaffolding to area of wall
20 m × 6 m high = 120 m²

Mats needed for this area: Standards (at 2 m c/s.) – 32 m; ledgers – 115 m; transomes – 50 m; bearers – 60 m; handrails – 60 m; bracing – 80 m; ties – 30 m; Total = 527 m Couplers and other fittings (say) 350no. Scaffold boards – 50no.

Purchase cost of above: Tubing 527 m at £3/m; fittings 350 at £4 ea; boards 50 at £10 ea. Total cost £3,481 + 2½% to cover loss and damage = £3,568

Allow (say) 40 uses therefore cost per use = £3,568/40 = £89

Labour cost: Erecting – allow 3 men × 1½ days = 36 man hours at £4/hr = £144
Dismantling – allow about half this time therefore cost = £72

Total lab. cost = £216

Transport to and from site: Weight of scaffolding:

tubing 527 m at 5 kg/m	2,635 kg
fittings 350no. at 2 kg each	700 kg
boards 50no. at 15 kg each	750 kg
	4085 kg

Total weight of 4 tonnes can be carried in one journey therefore allow for 2no. round trips (includes for delivery and collection) at £82 = £164

Maintenance; Allow 2hrs/week × 20 weeks at £4/hr. = £160

Total cost:	materials	£89
	labour	£216
	transport	£164
	maintenance	£160
		£629

Therefore cost per cm² of external scaffold = £629/120 m² = £5.24 (for full period scaffolding is required)

Total area of external face of building:

30 m × 7 m × 2	= 420 m²
7 m × 10 m × 2	= 140 m²
30 m × 3.5 m × 2	= 210 m²

770 m² at £5.24 = £4,035

Item B

Floor area 29.5 × 9.5 × 3 floors
= 841 m^2 at £1.00 per m^2 (based on trestles and boards) = £841

Item D

5no. at £45 each (for protection of work from weather) = £225

Item F

Incl. in item A.

Item G

Hire of portable traffic lights 4 weeks at £60 per week = £240

Scaffolding and gantries

		£
A	External scaffold	4,035
B	Internal scaffold	841
C	Fly screens (*N/A*)	--
D	Tarpaulins	225
E	Gantries (*N/A*)	--
F	Erection and dismantling scaffolding (*incl. in 'A'*)	--
G	Traffic control	240
	C/F to summary	5,341

Summary

		£
Transport	page 1	4,428
Plant and hoisting	2	7,305
Labour and materials installing and clearing plant	3	112
Site accommodation	4	9,497
Temporary services reinstatement, and sundries	5	4,268
Defects and extra labour costs	6	24,126
Cleaning site and clearing rubbish	7	6,812
Scaffolding and gantries	8	5,341
Carried forward to tender summary.		£61,889

Conclusion

The figure obtained for the Preliminaries can now be carried forward to the tender summary when all that remains to be done is the establishment of any tender adjustments where necessary and the overheads and profit.

Example 1 shows the Preliminaries Bill relating to the foregoing text and pricing schedule, with the rates previously calculated inserted against the appropriate bill items.

Example 1

BILL NO 1

PRELIMINARIES

	Preliminary particulars		£	p
A	Names of Parties	Employer Southshire County Council, County Hall, Southtown. Architect A. B. James, RIBA, County Hall, Southtown. Quantity Surveyor J. D. Hall & Partners, ARICS, Seaward Chambers, Southtown.		
B	Description of site	The site of the proposed works is at the Kingsway Shopping Precinct, Southtown. Access to and from the site is by way of the service road off Wall Road and indicated on the location plan. The extent of the site available to the Contractor, throughout the contract period for the purpose of the works, storage of materials and plant, erection of sheds etc. is within the temporary fence as indicated on the location plan.		

1/1 *To collection* £

PRELIMINARIES

£ p

The contractor shall restrict his men and any Sub-Contractor's workmen to within the temporary fence except for the purposes of site clearance, drainage works and erection of permanent boundary fences. Any damage caused through failure to comply with these restrictions shall be made good at the Contractor's expense.

The Contractor must examine the drawings and visit the site and acquaint himself with local conditions, nature of the soil, access to site, working and storage space, conditions affecting supply of labour and materials and the execution of the Works generally. No claim for extras will be admitted for errors or omission arising from the Contractor's failure to satisfy himself on these matters.

Any sand or gravel that may be discovered on the site during the excavation shall not be used for the Works.

A Description of Works

The Works comprise the erection of a Branch Library together with drainage, roads and pavings.

Drawings illustrating the Works are supplied on loan to the Tenderer and must be returned to the Architect by the date on which the tenders are due.

It is the intention of the Employer to enter into separate direct contracts for:

Site Layout and Planting

1/2 *To collection* £

			£	p
A	Submission of Tenders	Instructions regarding the submission of tenders are given on the form of tender. The Employer does not bind himself to accept the lowest or any tender and no remuneration or reward shall be made to any Tenderer for work entailed in submitting a tender.		

B *Definitions and qualifications*

Definitions — Words importing the singular only, also include the plural and vice-versa where the context requires.

The following abbreviations are used:-

BS	British Standard Specification.
BSCP	British Standard Code of Practice.
SMM	Standard Method of Measurement (6th edition Metric July 1978)

The following standard abbreviations in respect of metric units are used:-

mm. – Millimetre
m. – Metre
m^2 – Square metre
m^3 – Cubic metre
kN – Kilonewton
M N – Meganewton
kg – Kilogramme
No. – Number

The term 'the Works' shall mean the whole of the works envisaged by this contract, including unless expressly stated otherwise, Provisional Sums, the works of

1/3 *To collection* £ –

£ p

Nominated Sub-Contractors, Nominated Suppliers, Local Authorities and Public Undertakings.
Standard phraseology has been used for descriptions in these Bills of Quantities and the Tenderer's attention is drawn to the qualifications of the rules of the SMM which are given in various sections of BILL NO. 2.

A Quantities

Notwithstanding the provisions of SMM Clause A7, fractions of a unit or of a kilogramme less than half, which would cause an entire item to be eliminated have been regarded as whole units or whole kilogrammes.

B Cost Analysis Information

The Bills of Quantities have been prepared on an elemental basis and the elements into which the work is divided are listed in APPENDIX 1. The numbers in brackets within the description column indicates these elements. In certain cases the same measured item applies to more than one element and in this instance the item total has been divided into and inserted against its respective element number, i.e.

140 mm thick wall	(2.5)	102 m^2
	(2.7)	123 m^2

The foregoing example indicates that 102 m^2 of 140 mm thick wall are in element (2.5) and 123 m^2 are in element (2.7) (and the total area of 140 mm thick wall is 225 m^2). The Contractor is requested to extend each element total separately. The value of each

1/4 *To collection* £

			£ p
		element will be ascertained by the Quantity Surveyors for their private cost analysis records only.	
A	Pricing of Preliminaries	The Contractor is to note in pricing these Preliminaries he must enter in the column against the individual item any sum he shall so require and shall not enter any unallocated overall lump sum at the end of these Preliminaries.	
B	Rates in the Bills of Quantities	The individual rates in the Bills of Quantities are to be wholly inclusive of any percentage adjustment that the Contractor may desire especially upon Sub-Contractors' work and such percentage adjustment is not to be stated separately. In pricing the Bills of Quantities no alteration shall be made to the text, the descriptions given are to be rigidly adhered to and the items priced as described.	
C	*Contract particulars*		
	Conditions of Contract	The Conditions referred to in the Standard Form of Building Contract Local Authorities Edition with Quantities 1980 Edition as set out below, shall be applicable to any contract arising from this offer. A summary of alterations and deletions is given in APPENDIX.	
		The following words and expressions shall have the meanings hereby assigned to them except where the context otherwise requires.	

1/5 *To collection* £ ____________

PRELIMINARIES

£ p

(a) 'the Contractor' shall mean the person or persons, firm or company whose tender has been accepted by the Employer and including the Contractor's personal representatives, successors and permitted assigns.

Conditions of Contract (Contd)

(b) 'the Contract Drawings' and 'the Contract Bills' shall mean the Drawings listed in, and Bills of Quantities which accompanied the tender documents.

(c) 'the Contract Sum' shall mean the sum given in the Tender.

The Form of Contract will be the Standard Form of Building Contract Local Authorities Edition with Quantities 1980 issued by the Joint Contracts Tribunal and the prices in these Bills of Quantities will be deemed to cover the cost of complying with the Clauses contained therein as set out below. A summary of alterations and deletions is given in APPENDIX.

Schedule of Clause Headings

Clause no. 1	Interpretations, definitions etc.
Clause no. 2	Contractor's obligations
Clause no. 3	Contract sum – adjustments
Clause no. 4	Architect's/SO's instructions

1/6 *To collection* £ ________

			£	p
Schedule of Clause Headings (Contd)	Clause no. 5	Contract documents		
	Clause no. 6	Statutory obligations, notices and fees		
	Clause no. 7	Levels and setting out the works		
	Clause no. 8	Materials, goods and workmanship to conform to description etc. (*testing & samples*)	*149-00*	
	Clause no. 9	Royalties and patent rights		
	Clause no. 10	Person in charge		
	Clause no. 11	Access for Architect/SO		
	Clause no. 12	Clerk of works		
	Clause no. 13	Variations and provisional sums		
	Clause no. 14	Contract sum		
	Clause no. 15	VAT supplemental provisions		
	Clause no. 16	Materials and goods unfixed or off site		
	Clause no. 17	Practical completion and defects liability	*400-00*	
	Clause no. 18	Partial possession by Employer		
	Clause no. 19	Assignments and sub contracts		
	Clause no. 19A	Fair wages		
	Clause no. 20	Injury to persons and property and Employers liability		
		1/7 *To collection*	*£549-00*	

PRELIMINARIES

		£	p
Clause no. 21	Insurance against injury to persons and property		
Clause no. 22	Insurance of the works against clause 22 perils Clause 22A shall apply OR Clause 22B shall apply OR Clause 22C shall apply (see appendix)		
Clause no. 23	Date of possession and completion		
Clause no. 24	Damages for non completion		
Clause no. 25	Extension of time		
Clause no. 26	Loss and expense caused by matters materially affecting regular progress		
Clause no. 27	Determination by Employer		
Clause no. 28	Determination by Contractor		
Clause no. 29	Works by Employer		
Clause no. 30	Certificates and payments		
Clause no. 31	Finance (no. 2) Act 1975 Statutory tax deduction scheme		
Clause no. 32	Outbreak of hostilities		

1/8 *To collection* £

PRELIMINARIES

Schedule of Clause Headings (Contd)			£	p
	Clause no. 33	War damage		
	Clause no. 34	Antiquities		
	Clause no. 35	Nominated subcontractors		
	Clause no. 36	Nominated suppliers		
	Clause no. 37	Fluctuations Clause 38 shall apply OR		

Clause 39 shall apply
OR
Clause 40 shall apply

A Amendments and Additions to the Conditions of Contract to comply with the Standing Orders of the Council

The following amendments and additions shall be made to the Conditions of Contract, and shall be incorporated therein and shall form part of and be a material term thereof, and the Contractor shall allow in his tender for all costs and expenses involved in complying with these additional Conditions and Amendments:-

(i) Every contract which exceeds £25,000 in value or amount shall be in writing and under the seal of the Council, failing which no contract shall be deemed to have been entered into or become binding upon the Council. Every other contract, except a contract required in extreme urgency or a contract of a trivial character shall be in writing and shall be signed by the Chief Executive or other appropriate Officer, provided that a contract

1/9 *To collection* £ ________

£ p

required in extreme urgency shall subsequently be confirmed in writing by the Chief Executive or appropriate Officer. No member of the Council shall enter either orally or in writing into any contract on the Council's behalf. Every tender, specification and all Conditions of Contract or other document whatsoever which shall be intended to form the basis of any contract with the Council, shall include an express provision to the effect that the terms of this paragraph of this Standing Order shall form part of and be incorporated in the contract and be a material term thereof.

(ii) Where a contract exceeds £100,000 in value or amount and is either for the execution of works or for the supply of goods and materials and services otherwise than at one time, the Council shall require and shall take a guarantee bond sufficient for the due performance of the contract. For the purpose of tendering the Contractor shall assume that the guarantor will be bound in an amount representing 10 per cent of the contract sum.

(iii) All contracts where a specification issued by the British Standards Institution is current at the date of the tender and is appropriate

1/10 *To collection* £ –

Amendments and Additions to the Conditions of Contract to comply with the Standing Orders of the Council (Contd)

		£	p
	shall require that goods and materials used in their execution shall be in accordance with that specification.		
(iv)	In every written contract a clause shall be inserted to secure that the Council be entitled to cancel the contract and to recover from the Contractor the amount of any loss resulting from such cancellation if the Contractor shall have offered or given or agreed to give to any person any gift or consideration of any kind as an inducement or reward for doing or forehearing to do or for having done or foreborne to do any action in relation to the obtaining or execution of the contract or any other contract with the Council, or for showing or forebearing to show favour or disfavour to any person in relation to the contract or any other contract with the Council or if the like acts shall have been done by any person employed by him or acting on his behalf (whether with or without the knowledge of the Contractor), or if in relation to any contract with the Council the Contractor or any person employed by him or acting in his behalf shall have committed any offence under the Prevention of		

1/11 *To collection* £ ________

£ p

Corruption Acts 1889 to 1916, or shall have given any fee or reward the receipt of which is an offence under Subsection (2) of Section 177 of the Local Government Act 1972.

(v) The Council's fair wages clauses for the time being in accordance with the House of Commons Resolution shall be incorporated in and form part of every written contract for the execution of work or the supply of goods or materials and shall be observed by the Contractor accordingly. Every tender, specification and all conditions of contract or other document whatsoever which shall be intended to form the basis of any contract with the Council shall include a copy of such fair wages clauses as aforesaid and such clauses shall be a material term of any such contract.

Summary of alterations and deletions

See APPENDIX

Appendix to conditions of contract

See APPENDIX

1/12 *To collection* £ –

Appendix to conditions

			£	p
Defects liability period (if none other stated is 6 months from the day named in the certificate of practical completion)	17.2	________		
Insurance cover for any one occurrence or series of occurrences arising out of one event	21.1.1	£ ________		
Percentage to cover professional fees	22A	________		
Date of possession	23.1	*to be agreed*		
Date of completion	23.1	*13 months from date of possession*		
Liquidated and ascertained damages	24.2	*at the rate of £50 per week.*		
Period of delay	(a) 28.1.3.2	*3 months*		
	(b) 28.1.3.1, 28.1.3.3 to 28.1.3.7	*1 month.*		
Period of interim certificates (if none stated is one month)	30.1.3	________		
Retention percentage (if less than 5%)	30.4.1.1	________		
Period of final measurement and valuation (if none stated is 6 months from the day named in the certificate of practical completion)	30.6.1.2	________		

1/13 *To collection* £ ________

PRELIMINARIES

			£	p
Period for issue of final certificate (if none stated is 3 months)	30.8	______________		
Work reserved for Nominated subcontractors for which the Contractor desires to tender	35.2	______________		
Fluctuations	38 OR 39 OR 40	 *clause 39 to apply.*		
Percentage addition	38.7or 39.8	 *5%*		
Formula rules (where clause 40 is to apply)	40.1.1.1 rule 3 rule 3	 base month ________ non adjustable element ________ (not to exceed 10%)		
	rules 10 and 30 (1)	Part 1/Part 11 of section 2 of the formula rules is to apply		

1/14 *To collection* £ ______________

PRELIMINARIES

A	*Amplifications of conditions of contract*	£	p
Clause 7 Levels and setting out of the Works	The Contractor shall set out the work accurately to the true intent and meaning of the drawings and shall provide all necessary instruments, lines, apparatus and anything necessary for setting out. All profiles are to be firmly fixed in concrete and are to remain in position and be properly maintained during the period of the Works. No profile is to be removed without the permission of the Architect.		
	All figured dimensions are to be worked to in preference to scaling the drawings and all large scale drawings are to be worked to in preference to those of a small scale. All dimensions and levels are to be taken where necessary from the existing works and site.	*292-00*	
	The Contractor shall at the beginning of the contract provide a datum formed in concrete 150 mm × 150 mm in section and embedded at least 450 mm into the ground. This shall be placed in a position to be agreed with the Architect and shall form the datum from which all levels required on the site shall be taken. The datum shall be carefully preserved during the period of the contract and on completion shall be removed and cleared away.		

1/15 *To collection* *£292.00*

PRELIMINARIES

			£ p
A	Clause 10 Person-in-charge	The person in charge is not to be changed without the consent of the Architect.	*16,690-00*
B	Clause 14 Contract Bills	The Contractor shall provide the Architect with a copy of his priced Bill of Quantities within four days of it being requested. All items in the Bills of Quantities are to be priced in BLACK INK in detail and the Final Summary page signed. These shall be the contract Bills of Quantities and will be used for the purpose of the Contract. These Bills of Quantities comprise pages in numerical sequence and the Contractor shall check each page number immediately upon receipt of the documents and notify the Architect immediately if any pages are missing or duplicated or if any printed letter or figure is indistinct. No subsequent claim for loss, consequent upon the Contractor's failure to comply with this clause will be admitted. Bills No. 3 and 4 are marked 'provisional' and will be remeasured during the progress, and on completion of the works, by the Quantity Surveyor in accordance with Clause 13 of the Standard Form of Building Contract. The Contractor shall allow for providing such assistance as may be necessary for that purpose.	

1/16 *To Collection* *£16,690.00*

PRELIMINARIES

			£	p
A	Clause 16 Materials and Goods unfixed or off-site	Notwithstanding any remedies he may have under the contract against Sub-Contractors the Contractor is to be responsible for all materials and fittings delivered to the site for his own and all Nominated Sub-Contractor's use and shall make good any damaged or missing at his own expense and shall provide all necessary protection to such materials and fittings. This responsibility shall in no way be limited because of the early delivery of such materials or fittings.		
B	Clause 19 Assignment or sub-letting	Certain parts of the works are specified to be carried out by firms named in the Bills of Quantities. The General Contractor is reminded that the relationship with these Sub-Contractors is the same as with Sub-Contractors of his own choosing. The General Contractor should ensure that he maintains a firm control of all Sub-Contractors employed by him and to this end is expected to use the standard form of sub-contract issued by the National Federation of Building Trade Employers and Federation of Associations of Specialists and Sub-Contractors and approved by the Committee of Associations of Specialist Engineering Contractors. In dealing with Suppliers the Contractor is advised to use forms		

1/17 *To collection* £

		£ p
	of enquiry and sub-contract which will enable him to seek the same indemnity as the employer seeks from him.	
A Clause 21 Insurance against injury to persons and property	In addition to the insurance required by Sub-Clause 21.1 the Contractor is to provide for cover in respect of injury and damage to property real or personal, arising out of or in the course of or by reason of the carrying out of the Works and caused by any negligence, omission or default by the Contractor, his servants or agents or, as the case may be, of such Sub-Contractor his servants or agents.	*see Final Summary incl. in overheads*
	The amount of cover is to be £250,000 in respect of any one claim the number of incidents for which cover is provided being unlimited.	
B Clause 23 Possession, completion and postponement	Before commencing works on site the Contractor is to submit to the Architect a progress chart, which shall take the form of a Critical Path Analysis or such other form as may be approved by the Architect and be for the total contract period stated in APPENDIX. The contractor shall indicate on the chart the allowance he has made for time lost through inclement weather.	
C Clause 39 Fluctuations	For the purposes of this contract and the operation of Clause 39 the tender date shall be deemed to be	

1/18 *To collection* £ –

General matters £ p

A Items for Convenience in Pricing S.M.M. Clause B13

Plant, Tools and Vehicles
Provide all plant, tools and vehicles necessary for the proper execution of the Works.

The Contractor shall at all times take the necessary precautions to reduce noise to the minimum by the use of efficient silencers, mufflers for drills and all other means necessary in accordance with BS 5228 Code of Practice for noise control on construction and demolition sites.

4,428-00
7,305-00
112-00

B

Safety, Health and Welfare of Workpeople
Provide for all costs incurred by complying with all current Construction Regulations (1966) and the Health and Safety at Work Act 1974 appertaining to all workpeople (including those employed by Nominated Sub-Contractors) employed on the site. (*protective clothing*).

246-00

C

Site Management Costs
Provide for all on and off site management costs.

D

Labour-on-Costs
Provide for all costs in respect of all workmen for:-
(a) National Insurance Contributions
(b) Obligations under the Employment Protection (consolidation) Act 1978
(c) Sick pay or insurance in respect thereof
(d) Pensions
(e) Annual and Public Holidays

inc. in all-in rate.

1/19 *To collection* *£12,091.00*

PRELIMINARIES

		£ p
	(f) Travelling Time, Expenses, Fares and Transport	*6,595-00*
	(g) Guaranteed Time (*incl. in all-in rate*)	
Items for Convenience in Pricing S.M.M. Clause B13 (Contd)	(h) Non productive time and other expenses in connection with overtime	
	(i) Incentive and Bonus Payments	*incl. in all-in rate*
	(j) C.I.T.B. Levy	
	(k) Any other disbursements arising from employment of labour	
A	*Safeguarding the Works* Safeguard the Works, materials and plant including those of Nominated Sub-Contractors against damage or theft including all necessary watching and lighting for the security of the Works and the protection of the public.	
B	*Protection of Adjoining Property* The Contractor shall carry out the Works in such manner as not to cause annoyance or interference with the owners and tenants of adjacent properties.	
C	*Protection of Existing Buildings* The Contractor shall be responsible for any damage caused to any existing buildings and structures, etc., whether caused by his own workmen or those of his Sub-Contractors and shall make good at his own expense any such damage to the satisfaction of the Architect.	*1,119-00*
D	*Protection of Trees etc.* The Contractor shall be responsible for any damage to	

1/20 *To collection* £7,714.00

			£ p
		trees, bushes, etc., that are shown on the drawings or directed by the Architect to be retained whether caused by his own workmen or those of his Sub-Contractors. Adequate means of protection shall be provided where necessary to preserve such trees, etc., in their existing condition. Any reinstatement found to be necessary at completion due to any negligence of the Contractor or his Agent shall be at the Contractor's expense.	
A		*Maintenance of Roads, etc.* Maintain public and private roads, footpaths, kerbs, etc., off the site, and keep the approaches to the site clear of mud. The Contractor shall make good any damage caused by his own or any Sub-Contractor's or Supplier's transport at his own expense or pay all costs and charges in connection therewith. (*repairs to kerbs etc. and keeping roads clean*) (*traffic control*)	*528-00* *240-00*
B	Items for convenience in Pricing S.M.M. Clause B13 (Contd)	*Police Regulations* The Contractor shall at all times comply with any police regulation or restriction.	
C		*Protection of Waterways* Provide for taking all reasonable precautions to ensure the efficient protection of all streams and waterways against pollution arising out of or by reason of the execution of the Works.	
		1/21 *To collection*	*£768.00*

			£ p
	Obligations imposed by the employer		
A	Protection of Underground Services	The position of known underground apparatus relating to gas, water, electricity and telephone services and surface water and foul drainage are marked on the site plan. Although reasonable steps have been taken to ensure that site plan is as accurate as possible, the Contractor should not rely on its accuracy, nor should he assume that the information supplied is complete or exhaustive. He should take all such steps as may be appropriate to check that the information shown on the site plan is accurate and that there are no other services in existence which are not referred to on the site plan. These steps should include making enquiries of the statutory undertakers as to the position of their apparatus, the excavation of trial holes to confirm these positions, and an examination of the surface of the site.	
B	Prevention of Trespass	The Contractor should prevent trespass of workmen and other persons on to adjoining properties and buildings.	
C	Rights of Advertising	No advertisements shall be displayed without the written consent of the Architect, who shall have power to withhold such consent and shall have power to prescribe the form of any advertisement to which he consents. (*sign board*)	*292-00*

1/22 *To collection* *£292.00*

PRELIMINARIES

			£	p
A	Site Meetings	The Contractor, or his duly authorised representative, shall at the written request of the Architect attend such meetings as may be held by the Employer or the Architect to discuss the progress of the Works, and shall furnish at each meeting any information and particulars which may reasonably be required for such discussions.	–	
B	Work in Inclement Weather	It is the policy of the Authority to require contractors to maintain, whenever practicable, continuity of working and productivity during inclement weather (*incl. in all-in rate*)	–	
		The Contractor will be expected to avail himself of all reasonable means, such as artificial lighting, screens and shelters and aids to building in inclement weather, which are currently available in using his best endeavours to prevent or minimise any delays; and the extent to which he has done so will be taken into account when assessing any extensions of time which may be given under Clause 2 of the Form of Contract for frost, inclement weather and other like causes.	–	
C	Water for the Works	Provide clean, fresh water for use on the Works, pay all charges in connection therewith, provide all temporary storage, plumbing services, connections, etc., and clear away and make good on completion.	*See Final Summary* 702.00	

1/23 *To collection* £702.00

PRELIMINARIES

			£ p
		If connection is made to the existing building supply the Contractor will be required to have installed at his own expense a check meter to register the volume of water used for the Works and to allow a credit for the value of such water.	
A	Lighting and Power	Provide all electrical energy, including low voltage system for use on the works, pay all charges in connection therewith, provide all temporary connections, leads, fittings etc., and clear away on completion.	*1,190-00*
B	*Temporary works*		
	Temporary Works Generally	Clearing away temporary works and making good after shall be deemed to the included with the items.	–
		Notwithstanding S.M.M. clause B any re-seeding of areas will be the responsibility of the Employer. Reinstatement of top soil and sub-soil after clearance of temporary works and due to indiscriminate use of any plant will be the responsibility of the Contractor.	–
C	Temporary Roads	Provide and maintain temporary roadways of sleepers necessary for the proper complete and efficient running of the Works, alter same as required and remove on final completion of the contract and reinstate all groundwork and provide protection of public roads and kerbs.	*2,956-00*

1/24 *To collection* *£3,946.00*

PRELIMINARIES

			£ p
A	Temporary Buildings	Provide and maintain temporary office accommodation, messrooms, etc. for the use of the Contractor's staff and watertight sheds for the storage of materials, tools and tackle for the use of workmen employed on the site, in position to be agreed with the Architect. Provide and maintain a suitable temporary lockable office with a floor area of approximately 10 square metres together with separate latrine accommodation for the use of the Clerk of Works. The office is to be provided with a bench at least 750 mm wide and 2100 mm long with four plan drawers, four chairs, a two drawer lockable filing cabinet, bowl, soap, clean towels and adequate lighting, heating and attendance. Provide and maintain where directed separate temporary latrine accommodation for the use of the workmen including all necessary screening. Provide for any cost which may be incurred by the application of rates on temporary buildings.	*9,251-00*
B	Temporary Telephones	Provide and maintain a telephone service to the foreman's office (with an extension to the Clerk of Works office) for the full period of the Works and notwithstanding S.M.M. clause B.8.K the Contractor shall pay all charges and expenses in connection therewith.	*965-00*
		1/25 *To collection*	*£10,216.00*

PRELIMINARIES

			£ p
A	General Scaffolding	Provide all necessary temporary scaffolding for the proper execution and completion of the Works. An approximate gross floor area of 132 square metres has a finished ceiling over 3.50 metres but not exceeding 6.00 metres above the floor.	*4,876-00*
B	*Works by nominated sub-contractors*		
	Works by Nominated Sub-Contractors	The definition of general attendance on Nominated Sub-Contractors shall be as described in S.M.M. Clause B. The facilities referred to in S.M.M. Clause B. and which are deemed to be included in the items of general attendance on Nominated Sub-Contractors shall also be deemed to be included in the items of general attendance on Local Authorities and Public Undertakings. The Contractor shall obtain from all Sub-Contractors and other firms employed on the Works full particulars of their requirements with regard to chases, holes, recesses etc., and is to supply them with the necessary dimensions and other information required. If the Contractor fails to obtain the necessary accurate information any subsequent alterations required shall be at the Contractor's expense.	

1/26 *To collection* £4,876.00

		£	p
	The Contractor shall arrange with the Sub-Contractor his own programme for the carrying out of the Sub-Contract work.	–	
	The Architect reserves the right on behalf of the Employer to order and pay direct any of the Prime Cost Sums at his own discretion and deduct the Prime Cost Sum from the contract amount.		
	In the event of any Nominated Sub-Contractor being paid direct by the Employer because of the default of the Main Contractor under Clause or because of any Agreement made between the Main Contractor and Sub-Contractor or Supplier authorising payment direct then any cash discount that is due will be deducted and retained by the Employer.		
	The Contractor shall not order any expenditure of any Provisional or Prime Cost Sum without a written instruction from the Architect.		
A	*Goods and Materials from nominated suppliers*		
	The Clauses relating to Nominated Sub-Contractors shall apply equally to Nominated Suppliers wherever these are applicable.		
	The term 'Fix Only' shall be as described in S.M.M. Clause B.10 The Contractor shall be responsible for the ordering of items to be supplied by	–	

1/27 *To collection* £

			£ p
		Nominated Suppliers within the period of the fixed price quoted by the Supplier.	
A	*Fixing goods and materials supplied by the employer*		
		The term 'Fix Only' when used in connection with any material supplied by the Employer is to include unloading, checking, protecting, insuring against loss or damage by fire or theft, storing, distributing, hoisting in position and returning packaging materials carriage paid.	–
B	*Protecting drying and cleaning the works*		
	Protecting from the Weather	Provide for protecting the Works from inclement weather (*tarpaulins etc.*)	*225-00*
C	Drying the Works	A Provisional Sum is included in the Prime Cost and Provisional Sums section for temporary heating to control the humidity of the Works and to dry out the building.	
		The Contractor is to give notice to the Architect of the necessity for the provision of temporary heating and is to submit in writing his proposals for maintaining a temperature of 13° to 7° day and night and a quotation for carrying them out fourteen days before the heating is required.	–
D	Removing Rubbish and Cleaning	Clear up and remove all rubbish and surplus material as it accumulates and at completion,	

1/28 *To collection* *£225.00*

		£	p
	keeping the site clean, wash floors and pavings, clean glass both inside and out, clean out gutters and down pipes, oil ironmongery, touch up paintwork and deliver up the site to the Employer in a perfectly clean and tidy condition ready for occupation.	*2,352* *976*	
A *Contingencies*			
	A Provisional Sum is included in the Prime Cost and Provisional Sums section for Contingencies	–	

1/29 *To collection* *£3,318.00*

PRELIMINARIES

PRELIMINARIES

	£ p
Collection	
page 1/1	–
page 1/2	–
page 1/3	–
page 1/4	–
page 1/5	–
page 1/6	–
page 1/7	*549.00*
page 1/8	–
page 1/9	–
page 1/10	–
page 1/11	–
page 1/12	–
page 1/13	–
page 1/14	–
page 1/15	*292.00*
page 1/16	*16,690.00*
page 1/17	–
page 1/18	–
page 1/19	*12,091.00*
page 1/20	*7,714.00*
page 1/21	*768.00*
page 1/22	*292.00*
page 1/23	*702.00*
page 1/24	*4,146.00*
page 1/25	*10,216.00*
page 1/26	*4,876.00*
page 1/27	–
page 1/28	*225.00*
page 1/29	*3,328.00*
Bill no. 1 to Final summary	*£61,889.00*

Chapter 7

The calculation of unit rates for construction work

Whilst the value of certain aspects of the tender figure, such as Preliminaries, profit, overheads, adjustments in respect of fluctuation clauses, etc. are, to say the least, 'arbitrary'; the pricing of the individual work sections within the bill of quantities presents the estimator with a much more straightforward and 'mechanical' approach although the element of 'assumption' is far from being removed entirely! However from the bill descriptions he will know precisely what is required from the point of view of the work to be done and the type and quality of materials to be used, and if the bill has been prepared in accordance with the Standard Method, there will be no ambiguity or confusion regarding the nature or quantity of work to be carried out on that contract.

The unit rates to be calculated for the measured items in the bill will almost certainly be assessed on the basis of net cost initially, with profit and overheads being determined as a lump sum for the purpose of inclusion in the tender figure. It is only on being asked to submit his priced bill of quantities for checking that the estimator will decide whether to allocate a proportion of the overheads and profit figure to each bill rate or simply leave it as a lump sum in the Preliminaries section.

7.1 The composition of unit rates

The Standard Method, under its 'General rules' in Section A, states that individual items within the bill of quantities are deemed to include the following elements of cost:

(a) Labour and all costs in connection therewith.
(b) Materials, goods and all costs in connection therewith.
(c) Fitting and fixing materials and goods in position.
(d) Plant and all costs in connection therewith.
(e) Waste of materials.
(f) Square cutting.
(g) Establishment charges, overhead charges and profit.

On the assumption that item (g) above is excluded from the unit rate calculation at this stage, the above list may be condensed into the following:

1 Labour costs.
2 Material costs.
3 Plant costs.

As has been stated earlier, the estimator may choose to include his plant costs elsewhere within the tender sum, but for any measured item of work within the bill, the unit rate must contain one, two or all three of the above resources. Each element is normally priced separately in accordance with the stated unit of measurement for each item; the individual costs which are then added together represent the total anticipated net cost to the contractor of carrying out one unit of the work described.

The calculated price is then inserted in the unit rate column of the bill and this rate, which when multiplied by the quantity measured, produces the total cost of that operation on the contract which is then extended into the cash column in readiness for the final computation of the estimate at a later stage. This system helps management in the administration of the contract, if successful, as page totals for labour, materials and plant may be obtained for use in the preparation of bonus targets etc.

7.1.1 The labour element

The question of how much per hour skilled and unskilled labour is costing the contractor is established in the 'all-in' rate calculation. This hourly rate when multiplied by the amount of work (or output) the operative can complete in 1 hour will give the unit labour cost for the operation described.

In order to carry out the work contained in the bill items, the contractor may require the use of skilled or unskilled labour, or indeed a combination of both where the work described is to be tackled by a 'gang'.

Depending on the nature and extent of the work, a gang

strength may comprise 1 craftsman and 1 labourer (1 + 1), 2 craftsmen and 1 labourer (2 + 1), or 5 craftsmen and 2 labourers (5 + 2), etc. and entails the build up of the corresponding gang cost per hour based on the cost of each member of the team. Thus a 2 + 1 gang would cost (£4.50 + £4.50 + £4.00) = £13.00 per hour, where £4.50 and £4.00 are the all-in hourly rates for skilled and unskilled labour respectively. The labour rate per unit of work is then calculated on the basis of the amount of work which can be done in 1 hour by the gang as a whole.

The choice of how long it will take an operative to complete an item of work is crucial, and the decision in this regard must rest with the estimator who, no doubt working under pressure with time being limited, will have precious little time to give every bill item *too* much consideration. He must have, therefore, at his fingertips an appreciation of outputs for a wide range of building operations; to help him in this respect, he will have built up a library of 'constants'. The word 'constants', whilst in widespread use in the industry, is nonetheless very misleading since it implies a fixed rate of working. It is important, however, to remember that each contract is unique and as such outputs can vary considerably from one job to another according to the type of work, complexity, weather conditions, etc. Outputs cannot therefore be regarded as 'constant', and the stock of these which an estimator has to hand is recognised by him as being a guide only and the rates must be honed and finely tuned to suit the particular conditions applicable to the contract currently being priced.

7.1.2 The material element

The material content of the unit rate is somewhat easier to identify, since the specification and quantity are both stated in the bill of quantities and the estimator will have obtained from his prospective suppliers the current market price, although in many cases this will have to be converted and expressed in terms of the unit of measurement where this differs from the unit of purchase. Items for consideration when calculating the material element of a rate are:

(a) The cost of the material delivered to site.
(b) Unloading, handling and stacking cost.
(c) Waste.
(d) Subsidiary fixing materials.
(e) Returning of packing cases, crates and pallets, etc. (where applicable).

7.1.3 The plant element

The mechanical plant cost could be included in the unit rate only where it can clearly be seen that such costs are directly related to quantity – i.e. where an increase or decrease in the quantity of the bill item occurs perhaps following the issue of a variation order or the remeasurement of work, the corresponding mechanical plant cost increases or decreases in direct proportion to the adjustment in the quantity. In these circumstances the hourly, daily or weekly cost of the mechanical plant must again be converted and expressed in terms of cost per unit of measurement by establishing the performance or output of the machine in an average productive hour on site.

In many instances, however, it will be seen that mechanical plant costs are more closely related to time and method rather than quantity, in which case the estimator would probably choose to calculate costs in terms of lump sums for inclusion in the preliminaries or in the items provided at the beginning of the work sections.

Work priced as a lump sum

Where a contractor is identified as being the lowest tenderer, he will be required to submit his fully priced out bill of quantities for checking showing the breakdown of his tender into individual rates for each measured bill item. However, in the first instance as the estimator is in the process of considering the cost implications of the work he is pricing, 'unit rates', and even some of the measured bill items themselves may be largely irrelevant. As the estimator may prefer to think in terms of the total cost of performing some item of work which may incorporate a *number* of these measured bill items. For example, it may be more realistic to consider excavation costs in terms of the labour and mechanical plant requirements needed to complete the whole task within a given time-scale (information which would be readily available from the method statement and programme), in which case a single lump sum may be calculated in respect of a number of individual activities. Referring to the bill of quantities, the excavation work may cover several measured items such as reduced level, foundation, pit or basement excavation together with the corresponding disposal items of backfilling, removing surplus spoil from site or depositing in spoil heaps. Surface treatments and earthwork support items would also be measured and form part of the overall operation.

It is possible, therefore, that the draft priced bill of quantities will contain perhaps a number of large lump sums each covering two or more measured bill items, and it is only on being notified that his tender is the lowest that the estimator will have to turn his attention to the sometimes arbitrary allocation of these lump sums to the corresponding individual bill items, thus breaking them down into unit rates. It should be noted that this practice can present serious problems at post-contract stage where bill rates are being used to price variations, since it can be seen that bill rates do not always accurately represent the cost to the contractor of carrying out the corresponding operation described in the bill.

Work to existing buildings, refurbishment, alterations, etc.

In addition to pricing new work, it is likely that on many occasions the estimator will find himself pricing alteration, repair or rehabilitation work, which by its very nature presents him with further difficulties. Rehabilitation work involves restoring existing buildings to their previous or a proper condition, and having to work in and around existing buildings carrying out alteration and repair work poses additional problems for the contractor which must be given due consideration by the estimator.

Whilst his two principal objectives remain the same, i.e. ensuring that all anticipated costs are adequately covered whilst at the same time producing the lowest tender, the accurate identification of the nature and extent of the resources required can become something of a nightmare.

Labour

In connection with labour costs for carrying out rehabilitation work, the following items are worthy of attention when attempting to determine output:

1 Work is often difficult and executed in uneconomical 'small lots'.
2 Often, access is restricted, as are working hours when alterations and extensions are being carried out to a building in which staff are working who cannot be disturbed by noisy building operations. Overtime working may be required when the building is empty at weekends.
3 Removal of rubbish and debris may be difficult when working on upper floors with no direct access to the outside.
4 Existing furniture and equipment may have to be continually moved around.
5 It is likely that the nature of the work will preclude the use of many items of mechanical plant resulting in labour-intensive operations.
6 Extra payments are often required to be paid to operatives when working in difficult, dangerous and unpleasant situations in old ducts, boiler rooms, roof spaces, etc.

Materials

The following additional considerations should be borne in mind when calculating material costs:

1 Unloading, distributing and getting materials into position may present difficulties with a considerable amount of double handling and hoisting involved.
2 Problems may arise in trying to match up new materials with existing, e.g. skirtings, architraves, flooring, etc.
3 Materials are often purchased in small quantities.

Mechanical plant

Mechanical plant costs may be affected as follows:

1 Items of mechanical plant and equipment such as hoists and scaffolding, etc. may be subject to excessive periods of standing time.

Preliminaries

Facilities and services provided under the heading of Preliminaries may be affected as follows:

1 In connection with contractual obligations, the content and extent of the work may not be clearly defined with responsibility for existing defects and condition of the building being under dispute at a later stage.
2 Protection items may be costly with following trades working in close proximity to newly finished work.
3 Storage space may be restricted, resulting in goods and materials being stored off site or in an adjacent building.
4 Additional costs are often incurred in providing temporary services and the maintenance of existing services such as water and electricity for staff still working in parts of the building. Temporary dust screens, walkways and other protective measures must be priced for.
5 Security could be a problem in ensuring that the building is properly locked up and secured at the end of each working day.

Administration

Items to be considered include:

1 The remeasurement of large areas of work measured as provisional in the bill where the nature and extent of the requirements were uncertain when the bill of quantities was being prepared.
2 Supervision and co-ordination of the workforce, subcontractors, etc. becomes difficult as operatives are often working in isolated areas and rooms throughout the building and not in immediate contact with the foreman or chargehands.

The following work sections (7.2–7.11) providing examples of the method of pricing of typical operations to be found in the bill of quantities, those items to be priced being stated at the beginning of each work section. It should be noted, however, that all stated outputs can at best be described as an average guide only to performance and should therefore be treated with great caution. All-in rates of £4.50 and £4.00 per hour have been assumed for skilled and unskilled operatives respectively.

7.2 Excavation and earthwork

The nature of the work included in the excavation section of the bill of quantities falls broadly into six categories for the purpose of pricing by the estimator, namely:

1 *Site preparation work*. Involving cutting down existing trees, hedges, shrubs, etc., turfing and excavating top soil for future use.
2 *Excavation*. Covering various types of excavating operations.
3 *Earthwork support*. Involving the temporary support to sides of excavations where the estimator considers there is a need for this operation.
4 *Disposal of excavated material*. Involving the various methods available for disposing of any excavated material whether on or off the site.
5 *Filling*. Involving backfilling around excavations and making up levels generally in areas on the site using materials specified by the architect.
6 *Surface treatments*. Involving the preparation of the surface of the ground by compaction, blinding, etc.

7.2.1 Factors affecting the cost of excavation work

The pricing of the excavation work section is probably the most difficult and arbitrary of any since there are so many variables involved, each of which demand careful consideration.

The items which determine costs are as follows.

1. Weather conditions

Excavation work is obviously exposed to the weather, which cannot be predicted with any degree of accuracy even from day to day, and bearing in mind that the estimator will be preparing his price some weeks, perhaps months before the work is planned to start on site, an element of guesswork and a great deal of luck is needed in order for the excavation to be carried out profitably! In dry conditions, it can be expected that mechanical plant will perform well and achieve considerable savings in time and expense, but under wet conditions it tends to become bogged down and its efficiency seriously impaired.

2. Bulking

Excavation is measured 'net' in the bill of quantities, i.e. the quantity stated being that 'in the ground' **before** excavation. However, after being excavated, soil increases in volume (bulks) to a greater or lesser extent depending on the nature of the ground. For example, 1 m^3 of gravel or similar, when excavated, increases in volume to about 1.1 m^3 – a factor of 10 per cent, whilst clay bulks to a greater extent, 1 m^3 in the ground increasing to about 1.33 m^3 – a factor of 33⅓ per cent. Bulking factors for other types of ground are given below. This bulking factor affects the pricing of the 'disposal' items measured in the bill since a greater quantity of material will actually have to be moved than the corresponding 'measured' quantity in the bill.

Increases in bulk of materials after excavation

Type of ground	Increase in volume (%)
Gravel	10
Sand	12½
Ordinary earth	25
Clay	33⅓
Chalk	33⅓
Rock	50

3. *Type of excavation and volume*

The Standard Method requires the type of excavation to be described together with details of depth stages, etc. but does not dictate to the contractor the **method** of carrying out the work – the choice remains with him, although the method adopted will be one which results in the fastest rate of progress, given the circumstances, and at the lowest possible cost, which normally means the use of mechanical plant. However, a complex design, work in the vicinity of underground services, existing tree roots, etc. may all serve to restrict the use of machinery and demand the employment of hand digging, or maybe some combination of the two, a composite rate then being calculated.

Where the total volume of excavation is insufficient, it may not be economical to use mechanical plant since the cost of transporting it to and from the site may be too high in relation to the amount of work to be done, thereby precluding its use in favour of total hand digging, albeit costing far more per cubic metre than machine digging under normal conditions.

4. *The nature of the ground*

Mechanical plant is capable of excavating in all normal ground such as top soil, loamy soil, soft or sandy clay, soft chalk, etc., although the output will vary somewhat as indicated in the table on p. 251. However, output is further reduced if friable or 'soft' rock is encountered, and hard rock may take a considerable time to dig out and at great cost, with sufficient care being needed to avoid damaging the machine by undue vibration and jarring.

Where hand digging is involved, the nature of the ground will have a crucial bearing on output and if a large amount is envisaged, great care must be taken during the estimator's routine visit to the site to establish the exact nature of the prevailing ground conditions.

In addition to the effect the nature of the ground will have on the actual excavating operations, the need, or otherwise, for any earthwork support will also be dictated by the extent to which the face of the excavations can remain intact without collapsing. A loose, sandy soil may require continuous close boarding or steel sheet piling in order to support sides of trenches, etc. whilst a heavy clay soil may need no temporary support, being in itself sufficiently stiff.

5. Distance to a tip

The measured item in the bill for removing surplus spoil from the site will not specify the location of the tip. (Unless the client has directed otherwise in the preambles.) It is normally the contractor's responsibility to find a suitable tip and make all necessary arrangments with the owner regarding tipping fees, etc., although in many cases the district council or local authority may be engaged in land fill schemes and may welcome the tipping of suitable material on specified sites at no charge. Again, one of the main purposes of the estimator's early visit to the site would be to investigate such possibilities and, where a number of options exist, calculate the respective transport and tipping costs since the nearest tip may not always be the cheapest. Having chosen a suitable tip, the estimator must then select the correct type and number of transport vehicles needed to haul the excavated material to the tip in order to keep the excavators working continuously. The choice of vehicles will depend on the quantity of spoil to be removed from the site and the rate at which it can be excavated, although almost certainly lorries of one type or another would be used with dumpers being confined to short haul trips on the site where they are particularly suited to running on rough ground or exposed or wet sites which would be unsuitable for lorries. Where the contract involves demolition or excavation within existing buildings and space is severely limited, skips may have to be used to remove the rubbish – a convenient but very expensive method of disposal.

6. Banksmen

A 'banksman' is the title given to a labourer who is assigned to work with an excavator and whose task is to assist in setting out and guiding the machine along the right lines and to the correct depth, effectively being 'in attendance' whilst occasionally jumping down into the trench to dig out isolated sections of the work which may have fallen in behind the machine and awkward corners which cannot be reached by the excavators' bucket.

It is usual for the banksman, who it should be noted, is entitled to a small additional payment (see NWR 3), to be included in the cost of operating the excavator whilst assuming the work he does to be 'non-productive'.

The table opposite shows an approximate guide to the relative difficulty of digging by hand in various types of ground assuming 'ordinary' ground to have a value of 1.00. An output would be chosen based on excavating in ordinary ground which would then be multiplied by the 'multiplier' factor to obtain the corresponding rate in other conditions.

Adjustments for **hand** excavation in ground other than 'normal'

Type of ground	*Multiplier*
Sand	0.75
Stiff clay	1.25
Compact gravel	1.50
Soft chalk	2.50
Soft rock	4.00

7.2.2 Excavating equipment

Since the majority of excavation work is now carried out by machine, the main features of some of the more common items of excavating plant are worthy of a mention at this point.

Description of plant	*Main features*
Excavator with back actor and front-loading bucket	An extremely versatile machine available on wheels or tracks capable of trench, basement and foundation excavation with the added advantage of being able to load the spoil into lorries. The front bucket can also be used for spreading hardcore and for other general duties on site.
Face shovel	Particularly suited to working against banks forming cuttings and embankments and capable of excavating in almost any kind of ground. Unsuitable for work at ground level and being tracked, requires transport to and from the site.
Skimmer	Used for shallow excavation and surface stripping where large quantities justify its expense.
Grab	In addition to its use as an excavator, it is used for loading loose materials, being particularly suitable for depositing material in large spoil heaps on site. Its lack of mobility, however, is a disadvantage.
Dragline	Versatile machine for large-scale excavation below the level of its tracks.
Tractor and scraper	Economical only where very large areas of site stripping, levelling and grading are involved with material transported and deposited a relatively short distance.

7.2.3 Guide to average excavation outputs

(a) Hand excavation	
Operation	*Output per m^3 in unskilled hours*
Excavate to reduce levels	2.50
Excavate basement n.e.	
1 m deep	2.75
2 m deep	3.00
4 m deep	4.50
Excavate found. trench n.e.	
1 m deep	2.50
2 m deep	3.25
Wheeling excavated material (per 100 m)	1.00
Backfilling in layers and compacting	1.50
Loading excavated material into lorry	2.00

(b) Machine excavation			
Type of excavation	*Output of machine in m^3 per hour*		
	0.19 m^3 cap. bucket	0.29 m^3 cap. bucket	0.38 m^3 cap bucket
Site stripping	8	14	19
Reduce levels	9	15	21
Trenches	7	11	17
Basements	9	15	21
Pits	5	8	11

Adjustments for ground other than ordinary ground

Output in:

Stiff clay	× 0.82
Gravel	× 0.94
Firm sand	× 1.41
Soft rock	× 0.50

Increases in bulk of materials after excavation (%)

Gravel	10
Sand	$12\frac{1}{2}$
Ordinary earth	25
Clay	$33\frac{1}{3}$
Chalk	$33\frac{1}{3}$
Rock	50

Weights of various materials

Ordinary earth	0.62 m^3 per tonne

Sand	0.66 m^3 per tonne
Stiff clay	0.52 m^3 per tonne
Gravel	0.57 m^3 per tonne
Chalk	0.44 m^3 per tonne
Rock	0.37 m^3 per tonne

Earthwork support

Earthwork support is measured and priced in square metres to the actual face of the excavation which may require supporting. Often the estimator does not price this item or includes only a nominal amount if he feels the ground is sufficiently stable so as not to need any support. Should the estimator decide to price the earthwork support item and then subsequently the contractor chooses not to carry out the operation, the amount in the bill of quantities will still be paid since the sum represents the risk undertaken by the contractor in choosing to leave the sides of the excavation unsupported.

The cost of providing earthwork support in timber depends upon:

(a) The nature of the ground and depth of excavation.
(b) The type of excavation (basement, trench, pit, etc.).
(c) The cost of the timber to be used.
(d) The number of times the timber can be salvaged for re-use. (If the bill description calls for the earthwork support to be left in, then it will have only one use and its full cost must therefore be allowed.)

The average number of uses can be taken as being 10–15 depending on the quality of the timber and how carefully it is used. The operative fixing and stripping the earthwork support is called a 'timberman' and is paid a few pence more than the labourer's hourly rate in accordance with the NWRA.

An average output for a timberman fixing and subsequently stripping earthwork support after use can be taken as 0.10 m^3 per hour. The method of pricing is as follows:

(a) Prepare a suitable design for the earthwork support.
(b) Consider the cost of supporting a given length or area of trench, basement, pit etc.
(c) Calculate the cost of timber required, and divide by the assumed number of uses.
(d) Calculate the labour cost of fixing and stripping the timber to the length or area of excavation under consideration.
(e) Add the material and labour costs together and reduce the total to a rate per m^2.

Extra over excavation

Breaking out rock, concrete, brickwork, etc. is measured as 'extra over' the excavation in which it occurs. Where surface concrete, etc., in the form of paths or beds, is encountered the unit of measurement is the square metre, whereas that discovered under the ground and broken out in conjunction with bulk excavation is measured in cubic metres.

Since the item is measured 'extra over', the cost of excavating 1 m^2 or 1 m^3 in normal soil must be deducted from the total cost of breaking out as this would have previously been measured and priced in the bill of quantities.

Set out below are average outputs for breaking out rock, etc. based on compressor and 1 operator plus ½ labourer in attendance (for removing material) per m^3.

Material encountered in reduced level excavation:

Brickwork	0.7 hrs	Reinforced conc.	3.6 hrs
Concrete	1.5 hrs	Hard rock	2.9 hrs

Material encountered in trench excavation:

Brickwork	1.5 hrs	Reinforced conc.	7.2 hrs
Concrete	2.9 hrs	Hard rock	5.9 hrs

(*Note*: Operatives using pneumatic drills will be paid a few pence more than the labourer's hourly rate in accordance with the NWRA.)

Pipe trenches

The rate must include for the following operations and in accordance with clause D.13.8 of the Standard Method, the work is measured in metres.

(a) Excavation.
(b) Earthwork support.
(c) Grading bottoms.
(d) Backfilling and compacting.
(e) Disposal of surplus spoil.

The quantity and cost of each item must be calculated and then expressed in terms of cost per metre of trench.

Hardcore filling

The cost of placing hardcore in bulk filling or in beds can vary considerably depending on whether the delivery lorry can gain access directly into the final position of the material. Where this is not possible, the hardcore may have to be tipped some distance from the building and barrowed into place.

Hardcore filling over 250 mm thick is measured in cubic metres and under 250 mm thick in square metres, stating the thickness.

Surface treatments. Surface treatments are measured and priced in square metres either separately or included with the description of the hardcore. See Clause D.40.

Average outputs for barrowing and filling hardcore are:

Barrowing and filling:	bulk hardcore over 250 thick	1.2 m^3 per hr
	hardcore in beds n.e. 250 thick	0.8 m^3 per hr
Compacting with:	vibrating roller	0.4 hrs per m^2
	10 tonne roller	0.02 hrs per m^2

The quantity of hardcore measured in the bill of quantities is the amount *after* compaction, and since hardcore consolidates by approximately 20 per cent (depending on the type of stone), 25 per cent should be added to the basic cost of material to allow for this. 1 m^3 of hardcore weighs approximately 1.60 tonnes.

The typical bill items to be priced in this section are as follows:

					£	p
A	Cut down privet hedge 1.000 high and grub-up roots	150	m			
B	Excavate to reduce levels max. depth not exceeding 250 mm	210	m^3			
C	Excavate trenches to receive foundations starting at reduced level max. depth not exceeding 2.00 m and not exceeding 300 mm wide	325	m^3			
D	Extra over trench excavation for breaking out concrete *PROVISIONAL*	15	m^3			
E	Earthwork support to faces of excavation max. depth 2.00 m and not exceeding 2.00 m between opposing faces	330	m^2			
F	Allow for keeping excavations and surface of the site free of surface water	ITEM				
G	Remove excavated material from site	535	m^3			
H	Hardcore filling in beds 255 mm thick deposited and compacted in layers not exceeding 150 mm	620	m^2			
I	Level and blind surface of hardcore with 25 mm thick layer of sand	620	m^2			

The build-up of excavation rates

Item A – Cutting down hedges

Assumptions

1 The work will be tackled by a gang comprising 3 labourers cutting

down the hedging and grubbing up the roots by hand together with a lorry and driver removing the material to a tip nearby.

Basic prices:

All-in rate for: Labourer £4.00 per hour
Lorry and driver £13.00 per hour

Method

Hourly cost of gang: 3 labourers at £4.00 per hour = £12.00
1 lorry and driver at £13.00 per hour = £13.00
£25.00

Assume the 3 labourers with the drivers' assistance can clear and load 10 m of hedging in 1 hour.

$$\text{Cost per m} = \frac{£25.00}{10} = \underline{£2.50 \text{ per metre}}$$

Item B – Excavating to reduce levels

Assumptions:

1 The work will be carried out entirely by a $\frac{1}{2}$ m^3 excavator with a banksman in attendance.
2 The output of the excavator includes digging and loading directly into a waiting lorry.

Basic prices:

All-in rate for: Banksman £4.00 + 16p. (NWRA) = £4.16 per hour
Excavator and driver = £14.00 per hour

Method

Hourly cost of operation: Excavator at £14.00
Banksman at 4.16
= £18.16

Assume output of excavator is 15 m^3 per hour

$$\text{Cost per cubic metre} = \frac{£18.16}{15}$$
$$= \underline{£1.21 \text{ per } m^3}$$

Item C – Excavating trenches

Assumptions

1 The excavation will be carried out partly by machine, say 90 per cent, and partly by hand – 10 per cent.
2 A banksman is required to work with the excavator; this operation being priced separately from the hand digging, with a composite rate for both methods being established for insertion in the bill.

Basic prices:

All-in rates for labourers, banksman and excavator – as above.

Method

(a) *Machine digging*:

Hourly cost of operation as for item B – £18.16

Assume output of excavator when digging trenches falls to, say, 9 m^3 per hr

$$\text{Cost per } m^3 = \frac{£18.16}{9} = £2.02$$

(b) *Hand digging*:

Assume 1 labourer can dig 1 m^3 of trench in $2\frac{1}{2}$ hours

Cost per m^3 = £4.00 × $2\frac{1}{2}$ = £10.00

The cost of the respective methods can now be combined as follows:

Composite rate=	90% × £2.02	= £1.82
	10% × £10.00	= £1.00
Total rate per m^3		= £2.82

Item D – Breaking out concrete

Assumptions:

1 The concrete, probably old foundations, is encountered during the trench excavation and is to be broken out by a labourer operating a mechanical drill powered by a compressor, the time taken obviously depending on the state of the concrete.

2 The breaking out of the concrete is measured **extra over**, meaning that the quantity of the preceding trench excavation has not been adjusted to allow for the volume of concrete. In order to obtain the extra over cost of breaking out the concrete, therefore, the cost of the normal trench excavation must be deducted as shown below.

Basic prices:

All-in rate for labourer operating compressor £4.00 + 10p (NWRA)

Hire of compressor and tools £1.50 per hour

Method

Hourly cost of operation: Hire of compressor	= £1.50
Operator	= £4.10
	£5.60

Assume it takes 3 hours to break out 1 m^3 of concrete

Cost per cubic metre = £ 5.60 × 3

= £16.80

Deduct: Cost per cubic metre previously established for normal trench excavation in ordinary ground at £2.82 per m^3

Extra over cost of breaking out concrete = £16.80 − £2.82
= £13.98 per m^3

Item E – Earthwork support

Assumptions

1 Timber is to be used as the earthwork support in accordance *with the sketch below*.
2 The timber can be salvaged for re-use on future contracts thereby allowing the cost to be spread over a number of jobs – assume 10 uses on average.

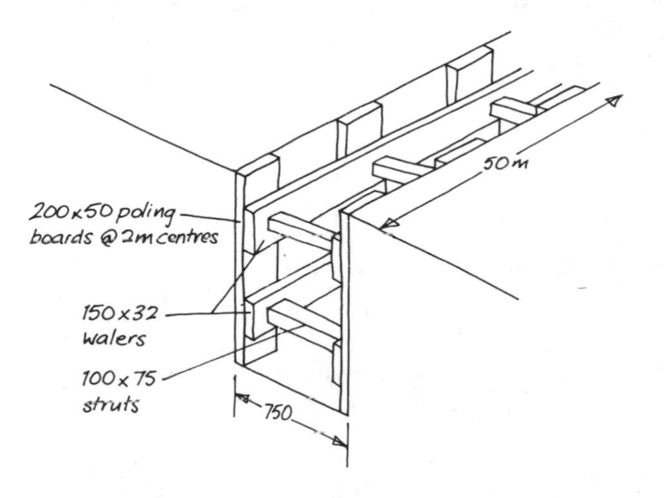

Basic prices:

All-in rate for labourer acting as 'timberman' £4.00 + 20p (NWRA)
Carcassing timber £210.00 per cubic metre.

Method

(a) Design the support system needed for the particular site conditions, assuming typical dimensions, sizes of timber, etc.
(b) Calculate the total quantity of timber needed for the design and establish its cost.
(c) Divide the total cost by the number of uses to obtain the cost per use.
(d) Calculate the **full** labour cost of fixing and stripping the above quantity of timber to obtain the labour cost per use.
(e) Add (c) and (d) to obtain the total cost per use.
(f) Divide (e) by the total area of excavation supported for this cost to obtain the cost per square metre.

(a) *Materials*:

Quantity of timber required for 50 m of trench:

Poling boards – number required = 50 m ÷ 2 m = 25 + 1 (end) = 26 × 2 (for both sides) = 52

52 × 1.00 × 0.20 × 0.05		= 0.52 m^3
Walers – 4 × 50 m × 0.15 m × 0.032 m		= 0.96 m^3
Struts – length of strut	= 0.75 m − 2(0.05 + 0.032)	
	= 0.75 − 0.168 m	
	= 0.582 m	
Quantity required	= 0.582 m × 52 × 0.10 m × 0.075 m	= 0.23 m^3
		1.71 m^3

Cost of carcassing timber delivered to site per m^3 =	£210.00
Allow 1 hour per m^3 unloading at £4.00 per hour =	4.00
	£214.00
Add 10% waste and cutting =	21.40
	£235.40 per m^3

Total cost of timber = £235.40 × 1.71 m^3 = £402.53

Allowing for 10 uses, cost per

use = $\frac{£402.53}{10}$ = £40.25 . . . [1]

(b) *Labour*:

Assume timberman can cut to size and fix 1 m^3 of timber in 20 hours – allow half this time for dismantling.

Total cost per m^3

= £4.20 × 30 hours = £126.00

Cost per use

= £126.00 × 1.71 m^3 = £215.46 . . . [2]

Total cost per use (1 + 2)

£40.25 + £215.46 = £255.71

Area of trench supported for this cost = 50 m × 1 m × 2 (both sides) = 100 m^2

∴ Rate per square metre = $\frac{£255.71}{100}$ = £2.56

Item F – Disposal of water

Assumptions

1 Two 100 mm dia. diaphragm pumps would be needed intermittently over a period of, say, 8 weeks whilst excavation is in progress to cope with surface water collecting in trenches, etc.

2 A labourer would spend, say, ½ hour per day in attendance on the pumps – cleaning, refuelling, etc.

Basic prices:

All-in rate for labourer attending on pump £4.00 + 9p (NWRA).
Hire of pumps £15.00 per week each.
Cost of fuel £3.00 per week each pump.

Method

Total cost to be calculated as a lump sum.

Hire of pumps: 2no. at £15.00 per week for 8 weeks	=	£240.00
Cost of fuel 2no. × £3.00 × 8 weeks	=	48.00
Cost of attendance: ½ hr per day × 40 days × £4.09	=	81.80
Total cost	=	£369.80

Item G – Removal of excavated material from site

Assumptions

1 The excavator is loading directly into 10 tonne lorries.
2 The cost of the excavator is included in the excavation rates.
3 Ideally, the optimum level of efficiency will be achieved when the number of lorries assigned to the task of disposing of the excavated material is such that the excavator is able to work continuously with an empty vehicle drawing in through the site gates as a previous one becomes fully loaded and is about to leave for the tip. In practice, this situation will be almost impossible to achieve. However, in theory this optimum number of vehicles can be calculated as follows:

$$\text{Number of vehicles} = \frac{\text{Time taken to load + Total journey time and tipping}}{\text{Time taken to load}}$$

Basic prices:

All-in rate for lorry and driver £13.00 per hour.
Tipping fees £5.00 per load.

Method

Assuming ordinary earth to weigh 1.60 tonnes per m³ – capacity of

$$\text{10 tonne lorry with full load} = \frac{\text{10 tonnes}}{1.60} = 6\tfrac{1}{4}\ \text{m}^3$$

Allowing for 25 per cent bulking, $6\frac{1}{4}$ m³ in the lorry is equivalent to $6\frac{1}{4} \times \frac{100}{125}\% = 5\ \text{m}^3$ 'in the ground'. In other words, for every $6\frac{1}{4}$ m³ of spoil actually removed, only 5 m³ will have been measured in the bill.

The cost of the lorry must therefore be based on the assumption that 5 m^3 only are removed each full load.

Where the removal of excavated material is sub-let to a haulage contractor, it must be remembered that his quotation will be based on *actual* quantities carted away irrespective of the quantity stated in the bill with the adjustment for bulking being made by the estimator. A quotation therefore of, say, £3.00 per m^3 would result in a bill rate of £3.00 $\times \frac{125}{100}\%$ = £3.75 where the bulking factor is 25 per cent.

Time taken to load 5 m^3 assuming an excavator output of 9 m^3 per hour = 60 min $\times \frac{5}{9}$ = 33 min.

Assume tip is 10 km from the site; total journey time is as follows:

Outward journey, say,	15 min
Tipping time, say	5 min
Return journey, say	12 min
	32 min

Therefore no. of lorries needed

$$= \frac{33 \text{ min} + 32 \text{ min}}{33 \text{ min}} = \frac{65}{33} = 2\text{no. (approx.)}$$

Total hourly cost of operation = 2 lorries at £13.00 per hour = £26.00

Since the excavator is assumed to be working continuously, the quantity of excavated material removed in 1 hour is 9 m^3

Cost per m^3 = $\frac{26.00}{9}$ = £2.89 . . . [1]

Add: Tipping charges at £5.00 per load.
Cost per m^3 assuming a full load to comprise 5 m^3 $= \frac{£5.00}{5}$
= £1.00 . . . [2]

Total rate ([1] + [2]) = £3.89 per m^3

Notes:

(a) It is unlikely that the above operation could be maintained since unavoidable delays in the lorries' journeys or problems encountered in the excavation work would result in either the lorries waiting for the excavator or the excavator waiting for lorries to return. It may therefore be advisable to make a more generous allowance for standing time thereby increasing the final cost per cubic metre.

(b) Where conditions prevent machine loading to take place, labourers may be called upon to carry out the work by hand. In this case the cost would increase considerably, reflecting the uneconomic methods of working, since at an assumed rate of loading of, say, 3 hours per m^3, it would take 15 man hours to fill the lorry!

Item H – Hardcore bed

Assumptions

1 The delivery lorry is able to back up to the excavations and tip the hardcore without any double handling.
2 The hardcore is spread around by hand and compacted with a hand-operated roller.
3 The hardcore is purchased by the tonne and weighs approx. 1.75 tonnes/m^3 (variable depending on the type of stone used).
4 Allowance for compaction. The quantity stated in the bill is the **net** area or volume of the void to be filled, and does not allow for a reduction in volume due to compaction. The estimator must therefore make the necessary allowance which, depending on the hardness of the stone, could be in the region of 20 per cent.

Basic prices:

Cost of hardcore delivered to site £8.00 per tonne.
All-in rate for: hand-operated roller £1.00 per hour.
operator £4.00 + 9p (NWRA)
labourer £4.00 per hour.

Method

(a) Materials:

Cost of hardcore d/d to site per tonne	=	£ 8.00
Cost per m^3 = £8.00 × 1.75	=	14.00
Allow 20% compaction	=	2.80
		£16.80
Add, say, 5% waste	=	0.84
		£17.64

Cost per m^2, 250 mm thick = £17.64 × 0.250 = £4.41 . . . [1]

(b) *Labour*:

(i) Spreading
Assume 1 labourer can lay 0.75 m^3 of hardcore per hour in beds

Cost per m^2, 250 mm thick

$$= \frac{£4.00 \times 0.250}{0.75} = £1.33 \quad \ldots [2]$$

(ii) Compacting

Hourly cost of operation: Hire of roller £1.00
Operator 4.09

£5.09

Since the hardcore is required to be compacted in layers not exceeding 150 mm, a 250 mm thick bed would need to be rolled **twice** to comply with this specification.
Assume 50 m^2 of hardcore can be compacted in 1 hour.

Cost per $m^2 = \frac{£5.09}{50}$

= 10p × 2 (for rolling twice) = 20p . . . [3]

Total rate ([1] + [2] + [3]) = £5.94 per m^2

Item I – Levelling and blinding

Assumptions

1 The operation is carried out by hand.

Basic prices:

Cost of sand delivered to site £9.00 per tonne.
All-in rate for labourer £4.00 per hour.

Method

(a) *Materials*:

Cost of sand delivered to site per tonne = £9.00.
Assuming the sand weighs approx. 1.50 tonnes per m^3.

Cost per m^3 = £9.00 × 1.50 = £13.50
Add 5% waste = 0.68

£14.18

The sand is simply spread around and therefore no allowance needs to be made for compaction.

Cost per m^2, 25 mm thick = £14.18 × 0.025 = 36p . . . [1]

(b) *Labour*:

Assume 1 labourer can blind 30 m^2 of hardcore surface with a 25 mm layer of sand in 1 hour.

Cost per $m^2 = \frac{£4.00}{30}$ = 13p . . . [2]

Total rate ([1] + [2]) = 49p per m^2

7.3 Concrete work

The nature of the work in the concrete work section of the bill of quantities falls broadly into six categories for the purpose of pricing, namely:

1 *In-situ concrete*. Covering both plain and reinforced concrete together with various labours and surface finishes which may be required.
2 *Reinforcement*. Covering various types of bar reinforcement which is priced by weight, and fabric reinforcement measured and priced by area or length depending on its location.
3 *Formwork*. To surfaces of the various types of *in-situ* concrete.
4 *Precast concrete*. Including lintels, copings, sills, etc.
5 *Prestressed concrete*. Including both precast and *in-situ* work.
6 *In-situ and precast flooring*. Including composite construction of *in-situ* and precast work and contractor-designed construction.

7.3.1 In-situ concrete (ready-mixed concrete)

Approximately 90 per cent of all *in-situ* concrete work performed on site is produced by the specialist ready-mixed concrete companies and a price per cubic metre can be readily obtained for the cost of transporting and placing the concrete. In many cases the transporting operation may be dispensed with if the ready-mix truck is able to manoeuvre into a position close to the foundations or whatever and empty its load directly into the trench or prepared formwork.

The price per cubic metre quoted by a ready-mix company will depend on:

1 The specification of the concrete.
2 The approximate total quantity which is to be ordered.
3 The distance which must be travelled by the delivery trucks from the ready-mix depot to the site.

Prices will normally be based on full loads (approx. 6.4. m^3) with surcharges being applicable on part loads.

The widespread use of ready-mixed concrete has been brought about by a number of reasons, some of which may be summarised as follows:

1 The lack of space available on many inner-city sites often means that site mixing is impossible since large open areas for the storage of aggregates and for mixing operations are required.
2 Where the manufacture of the concrete can be carried out on a large scale and under, as near as possible, ideal conditions, the quality of the finished product can be carefully controlled, whereas the quality of site-mixed concrete tends to vary to a greater or lesser extent with each mix.

3 The convenience of being able to obtain any quantity of high-quality ready-mixed concrete almost 'on demand' is usually well worth the small extra cost involved, especially since deliveries can be timed to meet planned production schedules, leading to a greater degree of efficiency.

7.3.2 In-situ concrete (site-mixed)

When pricing for site-mixed concrete, the rate calculation becomes more involved since **three** areas of cost must now be considered:

1 The cost of the materials required to make the concrete.
2 The cost of mixing involving both plant and labour.
3 The cost of transporting, placing and any surface treatments which may be specified.

1. The cost of materials

The constituent parts of concrete comprise cement and fine and coarse aggregates in varying proportions. The Code of Practice CP 110 sets out a table of prescribed mixes which shows the relationship between the various grades of concrete, representing characteristic strength, and the proportions of cement, fine aggregate and coarse aggregate needed to produce the various qualities and strengths of these concrete mixes. The grade of concrete will be included in the bill description of the item which will enable the estimator to consult the relevant table, perhaps reproduced in the trade preambles of the bill of quantities in order to determine the relative quantities of each material. In addition to determining the strength and workability of the concrete, the various proportions of cement, fine and coarse aggregate will also determine the cost, and since the cement is by far the most expensive ingredient, the higher the proportion of cement in the mix the higher the cost.

CP 110 expresses the proportions of the constituent parts in terms of the number of kilogrammes of dry material needed to produce 1 m^3 of fully compacted mixed concrete, the figures thereby taking into account the loss of volume due to shrinkage. It can be seen that the quantity of fine aggregate as a percentage by weight of the total dry aggregates varies according to the specification, but ranges between 30 and 50 per cent. For example, in order to produce 1 m^3 of concrete grade 25 using a maximum size of aggregate of 20 mm and giving a medium workability, the following quantities of dry materials would be needed:

Cement	=	360 kg
Fine aggregate (40% of 1,750 kg)	=	700 kg
Coarse aggregate (60% of 1,750 kg)	=	1,050 kg

Cement. The most widely used cement is Ordinary Portland Cement, which can be delivered to site in 50 kg bags for storage in watertight huts or purchased in bulk for subsequent storage in silos until required. The cost of unloading bagged cement will vary depending on the method of unloading and the distance from the point of unloading to the storage shed, although an allowance for hand off-loading of 1 hour per tonne should be sufficient.

Aggregates. Fine and coarse aggregates are purchased by weight or volume and delivered to site in tipper lorries, which obviates the need for an unloading operation since the materials can normally be tipped in heaps or bunkers in convenient positions close to the mixing area.

Water. A plentiful supply of clean water is required for the mixing process, but the cost of providing a temporary supply of building water is normally priced in the Preliminaries section of the bill or in the main summary.

2. *The cost of mixing*

Mixing will be carried out by one or more of the many different types of mechanical mixers available with only very small quantities being mixed by hand where output can be expected to be no more than 1 m^3 per 4 to 5 hours.

The number of men employed in the mixing operation will vary according to the size of mixer used – most small mixers up to 0.200/0.140 capacity are fed by hand shovel; however, with larger machines, hoppers can be fitted which will discharge cement and aggregates into the drum in accurate pre-determined proportions by means of gauge boxes or weigh batching equipment. An average gang would normally consist of two men; one mixer operator and one filler with the operator being entitled to an extra payment in accordance with the NWRA.

The choice of mixer. The size of mixer chosen for any job will depend on the means available of disposing of the mixed concrete, selection therefore being made on the basis of:

(a) The total amount of concrete required.
(b) The maximum required at any one time.
(c) The means of transport available.

Mixers are specified in accordance with their drum size, a .26/.20 mixer (Imperial notation – 10/7) being capable of accepting 0.26 m^3 of dry materials which after mixing shrinks to 0.20 m^3. This means that 0.20 m^3 of concrete is produced each mix and in order to keep the mixer working continuously at maximum output, say one mix every two minutes, sufficient means of transport must be available to cope with 0.20 m^3 $\times$ 30 mixes per hour = 6 m^3 per hour.

In practice, mixers of 0.20 m^3 capacity would normally only be used when wheelbarrows are being used for transportation, whereas a 0.40 m^3 mixer and over would be used in conjunction with a 0.5 m^3 skip or dumper. To use such equipment with a small mixer would mean that they would have to wait for 3 or 4 mixings before the skip or dumper would be full. Similarly, if wheelbarrows were to be used in conjunction with a large mixer, a disproportionate amount of time would be taken in emptying each mixing.

The cost of transporting and placing. The size and therefore output of the mixer will determine the number of men and barrows/dumpers needed to dispose of the concrete. A further determining factor is the distance the concrete has to be transported from the mixing point to its final position on site and the ease/difficulty with which it can be placed into position without segregation of the aggregates.

A concreting 'gang' will consist of:

1 or 2 operators/fillers (depends on size of mixer)
wheelers (depends on distance and rate of discharge)
spreaders (depends on final location)

Average mixer outputs in m^3 per hour

Rate of discharge in mins. per mix	Mixer size 7/5 (.20/.14)	10/7 (.28/.20)	14/10 (.40/.28)	18/12 (.51/.34)
2	4.25	5.95	8.49	10.19
2½	3.39	4.76	6.79	8.15
3	2.83	3.97	5.66	6.80
3½	2.43	3.39	4.85	5.83
4½	2.13	2.97	4.25	5.10
5	1.89	2.65	3.78	4.53
5½	1.70	2.38	3.39	4.07
6	1.41	1.99	2.83	3.40

Allow 1 hour per m^3 per 100 m round trip for wheeling and depositing concrete by hand.

Method of pricing

1 Determine output of mixer in m^3 per hour.
2 Calculate labour cost per hour for mixing, transporting and placing based on gang size.
3 Divide gang cost per hour by the mixer output in m^3 per hour to obtain the labour cost per m^3 of 'placed' concrete.

(Note: A labourer operating a mixer is paid a few pence extra per hour – see NWRA.)

Surface treatments The nature of the treatment to the surface of the concrete will be specified. For example, the surface may require power floating. The hire cost per hour of the machine chosen together with an operative must be expressed in terms of square metres and the bill item priced accordingly. Power floats are used on the surface of partially-set concrete and comprise three flat floats within a circular frame which helps to spread the weight of the machine and stops it sinking into the surface. The floats rotate and, as the machine is slowly moved over the surface of the concrete, a smooth polished face is produced. The hire rate of the float and operator would be divided by the area covered in one hour to obtain the rate per m^2.

7.3.3 Reinforcement

Reinforcement may consist of round steel bars, welded mesh fabrics or deformed bars. These materials are variously covered by BS 4449, 4461, 4466, 4482 and 4483.

The unit of measurement for bar reinforcement is the tonne, the diameter being given in the description (F. 11.2), whereas that for mesh reinforcement is the m^2 or m. depending on its location (F. 12.3 and F. 12.4).

Bar reinforcement

On most contracts, bar reinforcement is delivered to the site already cut, bent and labelled by the supplier in accordance with a 'bending schedule' supplied to him by the contractor, although on smaller jobs the contractor may choose to carry out the cutting and bending himself.

Waste. Where the contractor carries out the cutting and bending himself on site, a waste allowance of about 5 per cent is reasonable; however, if the steel is supplied in its final shape, an allowance of 1–2 per cent will normally be sufficient to cover losses and damage.

The weights of reinforcing bars in relation to their diameters are as follows:

Dia. in mm	*Weight in kg per m*
6	0.222
8	0.395
10	0.616
12	0.888
16	1.579
20	2.466
25	3.854
32	6.313

Fixing. Steel bar reinforcement is usually fixed by black tying wire at each intersection, the following being average quantities required per tonne of reinforcement:

6 mm–12 mm	10 kg
16 mm–25 mm	6 kg
32 mm–50 mm	4 kg

The fixing operation is normally carried out by skilled operatives who are paid a few pence per hour below the normal craftsman's rate (see NWRA).

The following table shows *average* labour outputs for reinforcement expressed in skilled hours per tonne for (a) cutting and bending and (b) fixing.

Location	*Diameter* 6		8		10		12	
	c&b	fix						
Beds	50	50	42	42	35	35	30	30
Floors, roofs, etc.	50	52	42	44	35	37	30	32
Walls	50	70	42	60	35	50	30	45
Beams, lintels, etc.	55	70	46	60	39	50	33	45
Links, stirrups, etc.	80	90	70	80	60	70	55	55

Location	*Diameter* 16		20		25		32	
	c&b	fix						
Beds	28	28	25	25	23	23	20	20
Floors, roofs. etc.	28	30	25	28	23	27	20	23
Walls	28	40	25	35	23	30	20	26
Beams, lintels, etc.	30	40	27	35	25	30	22	26
Links, stirrups, etc.	50	50	45	45	–	–	–	–

Unloading and stacking. For all sizes allow between 1 and 3 hours per tonne.

Spacers. Unless the bars are rigidly fixed in the correct position, the reinforcement may be displaced during concreting, particularly if the concrete is vibrated. In such cases, steel or plastic spacers may be used as support. It is sufficient to allow only a nominal cost for spaces per tonne of reinforcement fixed in position.

Mesh reinforcement

Mesh or fabric reinforcement is supplied in large sheets of various sizes, a typical size being 4.8 m × 2.4 m. It is specified by both its BS reference number and its mass per square metre, although a superficial or linear unit of measurement is used in the bill of quantities.

Waste. The normal amount of waste due to cutting, etc. will depend on its location in the works, typical allowances being as follows:

Beam wrapping	5%
Floors, slabs	$3\frac{3}{4}$%
Roads, beds	$2\frac{1}{2}$%

Laps. The quantity of mesh reinforcement is measured *net* in the bill of quantities; however, the material is required to be lapped by a minimum of 150 mm at the sides and ends and this 'loss' of area must be allowed for in the rate. The percentage addition for laps will depend upon the sheet size being used and the extent of the laps specified.

Example

Using the sheet size 4.8 m × 2.4 m with 150 mm side and end laps:

Actual area of sheet $= 4.8 \times 2.4 = 11.52\ m^2$

Net area covered $= (4.8 - 0.15) \times (2.4 - 0.15)$

$= 4.65 \times 2.25$

$= 10.46\ m^2$

Area of laps therefore $= 11.52 - 10.46 = 1.06\ m^2$

Percentage allowance for laps therefore $= \frac{1.06}{10.46} \times 100\%$

$= \underline{\text{approx. } 10\%}$

Fixing. Sheets of mesh reinforcement are usually fixed by unskilled labour and in most cases involves simply laying in position supported on pieces of broken brick, etc. where necessary; however the sheets, being large and cumbersome, require two men in order to manhandle them into place.

7.3.4 Formwork

The purpose of formwork is to contain freshly placed concrete until it has gained sufficient strength to be self-supporting; to produce a concrete member of the required shape and size; and to produce the desired finish to the surface of the concrete.

Types of formwork. Timber is the traditional material used for formwork, the facing material being either the timber itself or some other material such as plywood, steel, GRP, hardboard and expanded polystyrene. Several proprietary systems are also available such as slipforms, etc.

Design and support of formwork. The design and choice of materials to be used must be sufficient to resist all loads expected from the concrete, together with all live loads and self-weight of the forms themselves. Once a design has been chosen, the quantities of the various materials and support work used can be 'taken off' for pricing purposes.

Supports. Although timber could be used as props and supports, great use is now made of proprietary steel props and scaffold tubes. When allowing for the cost of these items, a hire rate per article will normally be given which necessitates an assumption of the length of time the items will be on hire. Typical minimum periods before formwork can be struck and props removed are as follows:

Props to slabs	11–14 days
Props to beams	15–21 days

Number of uses. Depending on the complexity of the design of the finished concrete work, the formwork used may be salvaged for re-use on some future contract. The number of times this can be done will vary from a single use for, say, the support to an *in-situ* single-flight concrete staircase, to perhaps 12 times where used in some straightforward, easily stripped location.

Before pricing any formwork, the tender drawings should be carefully examined in order to determine the design of each item and to assess the number of times each item can be used.

Method of pricing

(a) Materials.
  1 Calculate total quantity of materials needed for chosen design.
  2 Calculate total cost allowing for unloading, waste, etc.
  3 Allow a nominal figure for mould oil, nails and bolts.
  4 Divide total cost by the 'number of uses' to obtain the cost of materials **per use**.

(b) Labour involved in making.
  1 Allowing approx. 1.5 hours per square metre for a joiner, calculate the total labour cost of fabricating the item of formwork and again divide the cost by the 'number of uses' as above to obtain the cost of labour making **per use**.

(c) Labour fixing and stripping.
   1 Allow the **full cost** of fixing and stripping the formwork by a joiner which should include treatment of the shutter face, de-nailing, and repairs and cleaning. Again, the unit of cost will be **per use**.

Typical skilled labour outputs for fixing and stripping formwork per square metre

Location	*Fix*	*Strip*	
Horizontal soffits of floors, roofs, etc.	1.80	0.90	No 'making' involved with these items
Sloping soffits of floors, staircases, etc.	2.40	1.20	
Sloping soffits of ditto	2.70	1.35	
Extra per additional 1.5 m of strutting over 3.5 m high	0.24	0.12	
Extra on formwork to soffit of solid concrete floors or roofs over 225 mm thick for each additional 25 mm thickness	0.12	0.06	
Vertical or battering sides of foundations, etc.	1.50	0.75	
Ditto to sides of stanchion casings, columns, piers, pilasters, etc.	1.60	0.80	
Sides and soffits of horizontal beams, lintels	1.80	0.90	
Isolated beam casings and isolated beams	2.00	1.00	

(d) Hire cost of any props, column cramps, etc.
   1 Calculate total quantity of items needed from chosen design.
   2 Estimate the length of time the items will be needed on site.
   3 Calculate cost based on hire rate to give cost **per use**.

Add totals of (a), (b), (c) and (d), and divide by the surface area of concrete supported by the formwork to give the cost **per square metre**.

Where items are measured in **linear metres**, such as pilasters, beams, columns, edges of beds, roads, footpaths, etc., the same principle will apply except the total cost obtained should be divided by the **length** of concrete work supported.

The typical bill items to be priced in this section are as follows:

					£	p
A	Concrete grade 20 in foundation trenches 100–150 thick	110	m^3			
B	Concrete grade 25 in beds 150–300 thick	125	m^3			
C	Reinforced concrete grade 25 in beams, vibrated not exceeding 0.03 m^2 sect. area	35	m^3			
D	Power floating to surface of unset concrete	1,050	m^2			
E	Formwork to sides of column 300 × 350 sect. area	260	m			
F	Formwork to soffit of suspended slab	950	m^2			
G	12 mm diameter mild steel bar reinforcement in walls	2	tonne			
H	Ref. A142 fabric reinforcement weighing 2.22 kg per m^2 in ground slabs with 150 side and end laps (measured net)	1,050	m^2			

The build up of concrete work rates

Item A – In-situ concrete grade 20 in foundation trenches 100–150 thick

Assumptions

1 The concrete is mixed on site.
2 The mixed concrete is to be transported from the point of mixing to its position in the trench by dumper.

Basic prices:

All-in rate for: labourer £4.00 per hour
labourer driving dumper £4.00 + 12p (NWRA)
hire of dumper £1.20 per hour
hire of 0.40/0.28 mixer £1.40 per hour

Cement delivered to site in bags £68.00 per tonne
Fine aggregate £11.00 per tonne
Coarse aggregate £12.00 per tonne

Method

(a) *Materials*:
From the table of prescribed mixes as stated in CP 110, in order to obtain concrete grade 20 using 20 mm aggregate and with a fine aggregate content of 40 per cent, the proportions of

cement, fine and coarse aggregate in kg per cubic metre of mixed material allowing for shrinkage would be:

Cement	= 320 kg
Fine aggregate 40% × 1,800 kg	= 720 kg
Coarse aggregate (1,800 kg – 720 kg)	= 1,080 kg
Cost of cement delivered to site per tonne	= £68.00
Allow 1 hr per tonne unloading at £4.00	= 4.00
	£72.00

(Sand and gravel tipped – no unloading cost involved)

Material cost per cubic metre

320 kg cement at £72.00 per tonne	= £23.04	
720 kg fine aggregate at £11.00 per tonne	= £ 7.92	
1,080 kg coarse aggregate at £12.00 per tonne	= £12.96	
	£43.92	
Add 5 % waste	= £ 2.20	
	£46.12	[1]

(b) *Mixing and placing costs*:
Assume average rate of discharge from mixer is 6 mins per mix = 10 mixes per hour. Since the output of the mixer is 0.28 m^3 per mix, total quantity mixed per hour at this rate of working = 10 × 0.28 = 2.8 m^3.

Size and cost of gang required to transport and place:

1 mixer operator at £4.00 + 16 p (NWRA)	= £4.16
1 labourer filling at £4.00	= 4.00
1 dumper at £1.20 per hour	= 1.20
1 dumper driver at £4.12	= 4.12
hire of 14/10 mixer per hour	= 1.40
1 spreader at £4.00	= 4.00
	£18.88

Therefore in 1 hour, 2.8 m^3 of concrete can be mixed, transported and placed at a cost of £18.88.

Therefore cost per m^3 = $\frac{£18.88}{2.8}$ = £6.74 . . . [2]

Total rate ([1] + [2]) = £52.86 per m^3

Item B – Concrete grade C25 in beds 150–300 thick

Assumptions

1 Site mixed concrete as above.
2 The material cost per cubic metre will not be the same as item A since the proportions of the constituent parts are different.

3 The transporting and placing costs will differ from item A since it will take longer to place 1 m^3 of concrete in beds compared with foundations. Bearing this in mind, either the output of the mixer would have to be reduced, resulting in longer mixing times, or the gang strength would have to be increased to cope with the disposal of 2.8 m^3 which would be produced if the rate of output of the mixer was unchanged.

Method

(a) *Materials*:

400 kg cement at £72.00 per tonne	= £28.80
765 kg fine aggregate at £11.00 per tonne	= £ 8.42
935 kg coarse aggregate at £12.00	= £11.22
	£48.44
Add 5% waste	£ 2.42
	£50.86 per m^3 . . .[1]

(b) *Mixing and placing costs*:

Assume 1 extra spreader would be needed to supplement the gang strength established for item A.

Therefore hourly cost of operation = £18.88 + £4.00
= £22.88 per hr

Therefore cost per m^3 = $\frac{£22.88}{2.8}$ = £8.17 per m^3 . . . [2]

Total rate ([1] + [2]) = £59.03 per m^3

Item C – Reinforced concrete grade C25 in beams, vibrated

Assumptions

1 Ready-mixed concrete to be used, eliminating site-mixing costs.
2 The transporting and placing cost will depend on how near to the formwork for the beam the ready-mix lorry can get and on the method chosen for placing the concrete in position.
3 The cost of a hoist which would probably be needed to transport the concrete from ground level to upper floors would normally be included in the Preliminaries. The concrete to be carried in barrows.

Basic prices:

All-in rate for: labourer £4.00 per hour

operator of hoist £4.00 + 12p (NWRA)
vibrator operator £4.00 + 9p (NWRA)
Hire of vibrating poker £1.15 per hour
Ready-mixed concrete delivered to site £70.00 per m^3

Method

(a) *Materials*:

Cost of ready-mixed concrete delivered to site £70.00 per m³

Add 5% waste	£3.50	
	£73.50	. . . [1]

(b) *Placing cost*:

Hourly cost of operation:

1 hoist operator at £4.12	£4.12
1 vibrator poker operator at £4.09	4.09
1 spreader at £4.00	4.00
2 wheelers at £4.00	8.00
	£20.21 per hour

Assume that the gang is able to dispose of 4 m³ per hour

Therefore cost per m³ = $\frac{£20.21}{4}$ = £5.05 . . . [2]

Total rate ([1] + [2]) = £78.55 per m³

Item D – Power floating

Assumptions

1 The operation is carried out by a labourer using a power float.
2 The normal extra payments made under NWRA for working with mechanical plant are suspended for operations involving the use of power tools in order to provide a fine surface finish on concrete, with operatives receiving the equivalent of the normal craftsman's rate.

Basic prices:

All-in rate for: operative using power float £4.50 per hour
hire of power float £1.50 per hour

Method

Hourly cost of operation = £4.50 + £1.50 = £6.00

Assume an area of 10 m² can be floated in 1 hour

Therefore cost per m² = $\frac{£6.00}{10}$ = 60p

Formwork

Item E – Formwork to sides of column

Assumptions

1 Shutters made up of 150 × 25 tongued and grooved boarding nailed to 100 × 25 timber cleats secured by adjustable steel column cramps spaced at an average of 500 mm.

2 An allowance of 6 uses of the timber to be made with column forms supported by 100 × 75 timber.
3 A gang composed of craftsmen and labourers in the ratio 2:1 to be used to make, fix and strip the formwork.
4 Although it may be possible to strike the formwork earlier, it would be as well to allow a full week to take into account delays and bad weather, plus one further week's hire of clamps to cover erection and dismantling.
5 Whilst the unit of measurement in the bill is the metre, it is easier to visualise the cost of providing the formwork to the whole column as a lump sum initially which can then be reduced to a cost per m at the final stage.
6 The cost of the formwork can be based on the total height of the column not exceeding 3.5 m though in practice a drawing may be available thereby enabling the estimator to determine the actual height.

Basic prices:

All-in rates for labour as above
Sawn softwood carcassing timber £220.00 per m^3
150 × 25 tongued and grooved boarding £105.00 per 100 m
Hire of column clamps 50p per week per set

Method

(a) *Materials*:

Boarding – girth of column = 2 × 300 mm + 2 (350 + 25) mm allowing for laps
= 1.35 m

net width of boards = 150 mm − tongue (say 10 mm)
= 140 mm

number of boards needed = $\frac{1.350}{140}$ = 9.64

Therefore 10 boards must be allowed for.
Total length of boarding required for column
= 10no. × 3.5 m
= 35 m

Cleats – girth of cleat = 2 × 300 mm + 2 (350 + 50) allowing for laps
= 1.400 m

number of cleats required = $\frac{3.5 \text{ m}}{500 \text{ mm (spacing)}}$
= 8no.

Therefore total length of 100 × 25 for cleats
= 8no. × 1.4 m
= 11.2 m

Struts – assume 4no. required each 2 m long = 8 m

Total material cost per column

1 Boarding:

Cost of boarding delivered to site per 100 m	=	£105.00
Unloading and stacking $\frac{1}{4}$ hr per 100 m at £4.00	=	1.00
		£106.00
Add $7\frac{1}{2}$% waste		7.95
		£113.95

Therefore cost per col. $= \dfrac{£113.95 \times 35}{100}$

$= £39.88$

2 Cleats and struts:

Cost of carcassing timber delivered to site per m^3	=	£220.00
Allow 1 hour per m^3 unloading and stacking at £4.00	=	4.00
		£224.00
Add 5% waste		11.20
		£235.20 per m^3

Total quantity of timber required =

11.2 m × 0.1 × 0.025	=	0.028 m^3
8 m × 0.1 × 0.075	=	0.060 m^3
		0.088 m^3

Therefore cost per col. = £235.20 × 0.088 m^3

= £20.70

Total cost of materials per column = £39.88 + £20.70

= £60.58

Allowing for 6 uses, cost per use $= \dfrac{£60.58}{6}$

= £10.10

Allow for mould oil, nails, bolts (say) 0.50

£10.60 per use . . . [1]

(b) *Hire cost of clamps*

8 sets required for column at 50p per set per week × 2 weeks

= 8no. × 50p × 2 weeks

= £8.00 per use . . . [2]

(c) *Labour*:

Assume 1 craftsman can make, fix and strip formwork to one column in 6 hours with a labourer in attendance 50 per cent of the time.

Therefore total cost =	8 hrs at £4.50	=	£36.00
	4 hrs at £4.00	=	16.00
			£52.00 per use. . . [3]

Total cost per use ([1] + [2] + [3]) = £70.60

Therefore cost per m = $\frac{£70.60}{3.5 \text{ m}}$ = £20.17

Item F – Formwork to soffit of suspended slab

Assumptions

1 The design of the formwork comprises shutters of 25 mm thick external quality plywood decking on 150 × 50 joists at 600 mm centres supported at say 2 m centres by 150 × 50 ledgers, the whole structure being supported by adjustable steel props at, again, 2 m centres.
2 Allow 2 weeks' period before striking the formwork plus a further 2 weeks' hire period of props to cover erection and dismantling.
3 Assume the area of suspended slab to be 10 m × 10 m = 100 m².
4 Gang strength ratio as before.
5 Allow 7 uses for the formwork.

Basic prices:

All-in rates for labour as before.
Carcassing timber as before.
25 mm external quality plywood £7.00 per m².
Hire of adjustable steel props 60p per week each.

Method

(a) *Materials*:

Plywood – Cost of 25 mm plywood delivered to site per m²	= £7.00
Allow for unloading, say	0.17
	£7.17
Add 5% waste	0.36
	£7.53 per m²

Joists – Quantity required = $\frac{10 \text{ m}}{0.6}$ = 17 + 1 (end)
= 18 × 10 m = 180 m

Ledgers – Quantity required = $\frac{10 \text{ m}}{2}$ = 5 + 1 (end)
= 6 × 10 m
= 60 m

Total length of timber required = 240 m

Therefore total quantity = 240 m × 0.15 × 0.05 = 1.8 m^3
Therefore total cost = £234.15 × 1.8
= £421.47

Therefore cost per m^2 = $\frac{£421.47}{100}$
= £4.22

Total material cost = £7.53 + £4.22 = £11.75

Allowing 7 uses, cost per use = $\frac{£11.75}{7}$
= £1.68

Allow for mould oil, nails, etc. = 0.50

£2.18 per m^2 . . . [1]

(b) *Hire of props*:
Number required at 2 m centres = 6 × 6 = 36no.
36no. × 60p × 4 weeks hire = £86.40 per use

Therefore cost per m^2 = $\frac{£86.40}{100}$
= 86p . . . [2]

(c) *Labour*:
Assume gang strength of 4 craftsmen and 2 labourers a total of 4 days for making, erecting and striking the formwork to an area of 100 m^2

Total cost = 4 × 32 hrs × £4.50 = £576.00
2 × 32 hrs × £4.00 = 256.00

£832.00

Therefore cost per m^2 = $\frac{£832.00}{100}$ = £8.32 . . . [3]

Total rate ([1] + [2] + [3]) = £11.36 per m^2

Reinforcement

Item G – 12 mm bar reinforcement in walls

Assumptions

1 The reinforcement is delivered to site cut, bent and labelled.
2 The reinforcement is fixed by a qualified steelfixer who would be entitled to the normal craftsman's rate.

Basic prices:

12 mm bar reinforcement £500.00 per tonne
Tying wire £3.50 per kg
All-in rate for: labourer £4.00 per hour
steelfixer £4.50 per hour

Method

(a) *Materials*:

Cost of 12 mm dia. reinforcement delivered to site per tonne = £500.00

Cost of 12 mm dia. reinforcement delivered to site per tonne	= £500.00
Allow 2 hrs per tonne unloading at £4.00	8.00
	£508.00
Allow 10 kg of tying wire per tonne at £3.50 kg	35.00
Allow for plastic spacers, say	10.00
	£553.00
Add 2% waste	11.06
	£564.06 . . . [1]

(b) *Labour*:

Assume 1 steelfixer can fix in position and tie 1 tonne of 12 mm dia. reinforcement in walls in 45 hrs.

Therefore cost per tonne = 45 hrs × £4.50 = £202.50 . . . [2]

Total rate ([1] + [2]) = £766.56 per tonne

Item E – Ref. A142 mesh reinforcement in ground slabs

Assumptions

1 The reinforcement is supplied in sheets size 4.8 m × 2.4 m.
2 The sheets of fabric reinforcement would normally be laid by labourers.

Basic prices:

A 142 mesh reinforcement £2.00 per m^2

Method

(a) *Materials*:

Cost of fabric reinforcement delivered to site	= £2.00 per m^2
Allow for unloading, say	0.05
	£2.05
Allow for laps and waste, say 15%	0.31
Allow for tying wire, spacers, say	0.05
	£2.41 per m^2 [1]

(b) *Labour*:

Assume 2 labourers can fix 1 sheet (10.5 m^2 net) in 10 mins.

Therefore cost per $m^2 = \dfrac{0.33 \text{ hr} \times £4.00}{10.5}$ = 13p per m^2 . . . [2]

Total rate ([1] + [2]) = £2.54 per m^2

7.4 Brickwork and blockwork

The work measured in this section of the bill of quantities can be divided into six broad categories, namely:

1 Common brickwork in various locations.
2 Brickwork which is required to have a 'fair face' and may include work which is built entirely of facings, and common brickwork which has facings on the 'surface' only and which is known as 'facework'.
3 Brickwork, usually engineering bricks or similar in connection with boilers.
4 Blockwork of whatever material.
5 Damp-proof courses.
6 Sundries such as cutting holes and chases, bedding of frames and plates, flues, centering, etc.

7.4.1 Brickwork

The cost of brickwork per square metre in any class of work will comprise:

1 The cost of the bricks.
2 The cost of the mortar.
3 The cost of labour required to build one square metre.

1. The cost of the bricks

Bricks are normally purchased by the thousand and a cost per thousand based on full or part loads will be obtained from a supplier, usually including delivery to site. On arrival at the site, the bricks can be: (a) tipped; (b) hand off-loaded; or (c) crane off-loaded if the delivery lorry is equipped with a crane and the bricks 'strap packed'.

Bricks would be tipped only if they were fairly cheap commons subsequently covered by plaster, or in substructure where chipped and damaged bricks would not be seen. However, where engineering and facing bricks are concerned, careful unloading and stacking is essential not only because of their high cost, but also because of the likelihood of their rejection by the clerk of works where the face may be chipped rendering them fit for walling in only as commons out of sight. Crane off-loading is often available at an extra cost and this needs to be taken into account when pricing.

Waste. According to the results of a recent BRE site survey, the average wastage of bricks due to damage, cutting, over-ordering, theft, etc. was found to be:

Facings	12%
Commons	8%
Engineering	6%

These figures should of course be taken as a guide only as actual waste levels will vary considerably from one site to another.

Number of bricks per square metre. A standard brick measures 215 mm × 102.5 mm × 65 mm although a number of other sizes are available. The number of bricks required per square metre will therefore vary according to the size, but for the 'standard' brick there are 59 required per square metre per half brick thickness of wall allowing for a 10 mm joint.

2. The cost of mortar

Brickwork and blockwork is normally built in cement or cement-lime mortar depending on the location of the work, the mix being specified in the bill description. Mortar can be mixed on site or delivered to site ready-mixed and coloured as required in the form of a mixture of lime and sand to which cement only needs to be added to produce the mortar ready for use.

An allowance of approximately 33⅓ per cent needs to be added to the cost of the dry materials before mixing to cover for both shrinkage and waste. The quantity of mixed mortar required per square metre of brickwork will also vary depending on whether the bricks have one, two or no 'frogs' and also on the thickness of the wall. The quantity can be found by deducting the face area of 59 bricks from 1 m^2, which will give the face area of mortar required per square metre of brickwork, which when multiplied by the thickness of the wall will produce the quantity in cubic metres, remembering to make due allowance for the required number of vertical wall joints.

Conversion factors for cement, lime and sand are (tonnes per cubic metre):

Cement	1.44
Lime	0.60
Sand	1.52

3. The labour cost

Brickwork is normally carried out in gangs comprising 2 bricklayers and 1 labourer, although 5 + 2 gangs or other combinations may be employed on larger jobs. The labourer is assumed to be entirely non-productive when considering the actual laying of the bricks since his task is solely to keep the 2 bricklayers constantly supplied with bricks and mortar. The labourer engaged on this operation will not usually be involved in mixing the mortar, the labour cost for this operation being included in the build-up of the cost of mortar.

Type of wall	Average output per skilled hour
Half-brick wall in commons	55
One-brick wall in commons	60
One-and-a-half brick wall in commons	65
Brickwork in projections	50
Facing brickwork generally (incl. pointing as work proceeds)	40
Engineering brickwork (incl. pointing as work proceeds)	45

Average outputs for sundry work	Skilled hrs	Unskilled hrs
Extra for fair face and pointing	0.30 per m^2	0.15 per m^2
Form 50 mm wide cavity	0.10 per m^2	0.05 per m^2
Close 50 mm cavity at openings	0.35 per m	0.18 per m
Extra for eaves filling	0.40 per m	0.20 per m
Bonding half-brick wall to existing	1.10 per m	0.55 per m

Brick facework

Where brick facework comprises facing bricks with a common brickwork backing, the work is measured **extra over** which means that the area is measured firstly as though the wall were built entirely of commons and then again as 'extra over' for facings. The unit rate for this type of work must therefore take this into account – the cost of building a square metre of the wall in commons being **deducted** from the cost of building the same square metre in facings.

The number of facing bricks required per square metre of brickwork to a wall which is faced one side only is as follows:

Stretcher bond	59
English bond	89
Flemish bond	79
English garden wall	74 (3 courses of stretchers – 1 of headers)
Flemish garden wall	67 (3 stretchers and 1 header in the same course)

7.4.2 Blockwork

There are a large number of manufacturers of proprietary lightweight insulating blocks of various materials including foamed slag, concrete, clay, etc. in a wide range of strengths, thicknesses and face sizes with surfaces suitable for a fair smooth finish ready for direct decoration or keyed for subsequent covering with plaster. A typical face size of blocks would be 440 × 215 mm which when built with a normal 10 mm joint produces a 'walled-in' size of 450 × 225 mm allowing integration with standard brick courses (i.e. 65 + 10 mm = 75 mm) or three brick courses to each course of blocks).

Block thicknesses range from 60 to 215 mm in the following typical stages: 60, 75, 100, 140, 190, and 215 mm.

Unlike bricks, blocks are normally purchased by the square metre with the same mechanical off-loading facilities being provided in most cases. With a face size of 440 × 215 mm, the number of blocks required per m^2 of blockwork allowing for a 10 mm joint would be:

Area of 1 block = $(440 + 10) \times (215 + 10) = 0.450 \times 0.225$
$= 0.1013\ m^2$

Therefore number required per m^2 $= \dfrac{1}{0.1013}$
$= 9.87$no.

and the quantity of mortar required per m^2 for 100 mm thick wall would be:

Area of mortar on face of blockwork $= 1 - (9.87 \times 0.440 \times 0.215)$
$= 1 - 0.934$
$= 0.066\ m^2$

which, when multiplied by the thickness of the wall,
$= 0.066 \times 0.100$
$= \underline{0.0066\ m^3}$

Waste. An average waste allowance may be taken as being approximately 5 per cent since, although lightweight blocks are easily damaged in handling, a large number of half blocks are always needed for work at reveals, etc., thereby allowing damaged blocks to be utilised.

Labour. As with brickwork, labour is normally employed in the ratio of 2:1, approximate outputs being as follows:

Type of wall	***Skilled** hours per m^2*
100 mm lightweight blockwork in walls or partitions	0.50
140 mm as above	0.65
215 mm as above	0.90
Allow approx. 30 per cent extra labour for dense load-bearing concrete blockwork.	
Fair face and pointing one side as work proceeds	0.20

7.4.3 Sundries

Damp-proof courses

With the exception of brick and slate, d.p.c.s are supplied in rolls in a variety of widths to suit brick and block dimensions, the most common materials being polythene and hessian-based bituminous felt which are simply rolled out horizontally on top of the selected course of brickwork and bedded in the brickwork mortar ensuring that the specified laps at the ends of rolls have been provided. The waste factor for d.p.c.s is very small, with 2½–5 per cent adequate in most cases even after making due allowance for laps – the cost of the bedding mortar will have been included in the item of brickwork priced earlier and therefore no further allowance needs to be made here.

Holes and chases

Often the holes and chases required as 'builders' work in connection with services cannot be prepared at the time the bulk of the brickwork is being carried out as the positions may not at that stage have been verified by the trades concerned, which means that the marking and cutting of holes, ducts, etc. must be carried out later by hand or drill at greater expense than would have been otherwise. Allowance needs to be made for this when assessing the time involved in such operations.

Typical outputs allowing for 50 per cent labourer in attendance are as follows:

Type of work	*Skilled hours*	*Unskilled hours*
112 wide polythene d.p.c. laid horiz. per 30 m roll	0.25 per roll	0.12 per roll
As above, fixed vertically	0.37 per roll	0.18 per roll
Prepare top of old 1 brick thick wall for raising	0.25 per m	0.12 per m
Bedding wood frame in mortar	0.05 per m	0.03 per m
Extra for pointing one side in mortar	0.06 per m	0.03 per m
Hole for small pipe through ½B wall in commons	0.15 each	0.075 each
As above, in facings, incl. making good	0.30 each	0.15 each
As above in 100 mm blockwork	0.15 each	0.075 each

The typical bill items to be priced in this section are as follows:

					£	p
A	One brick thick wall in common bricks in English bond in cement mortar (1 : 3)	500	m^2			
B	65 mm facing bricks in skin of hollow wall half-brick thick in stretcher bond in cement-lime mortar (1 : 2 : 9)pointed with a neat flush joint as work proceeds	725	m^2			
C	Extra over common brickwork for facings in English bond in cement-lime mortar (1 : 2 : 9) pointed with a neat flush joint as work proceeds	50	m^2			
D	Form cavity 50 wide in hollow wall with 4no. wall ties per m^2	725	m^2			
E	Close 50 wide cavity at jambs with common brickwork	360	m			
F	112 wide polythene damp-proof course laid horizontally (measured net)	240	m			
G	100 thick concrete block wall in cement-lime mortar (1 : 2 : 9) finished fair one side	1,070	m^2			
H	Hole for small pipe through half-brick thick wall in facings and make good one side	80	no.			

The build up of brickwork and blockwork rates

Item A – One-brick thick wall in commons in cement mortar (1 : 3)
Basic prices:

All-in rates for labour as before
Common bricks £100.00 per thousand
Cement £68.00 per tonne
Building sand £11.00 per tonne
All-in hire rate for 0.15/0.10 mixer £1.40 per hour

Method

(a) *Materials*:

Bricks – Cost of bricks delivered to site per 1,000 = £100.00
(no unloading – assume commons are tipped) = 8.00
Add 8 per cent waste £108.00

Number of bricks required per m^2 = 118

Therefore cost per m^2 = $\frac{£108.00 \times 118}{1{,}000}$ = £12.74 . . . [1]

Mortar – Cost of cement delivered to site £68.00 per tonne
Allow 1 hr per tonne unloading £4.00
£72.00

Cost of (1 : 3) mix:
1 m^3 of cement costs £72.00 × 1.44 = £103.68
3 m^3 of sand cost £11.00 × 1.52 × 3 = 50.16
£153.84
Add 33⅓% for shrinkage and waste 51.23
Cost per 4 m^3 = £205.07

Therefore cost per m^3 = $\frac{£205.07}{4}$ = £51.27

Cost of mixing:
All-in rate for mixer = £1.40 per hour
All-in rate for operator = £4.00 + 9p = £4.09
(NWRA) £5.49 per hour

Assume output of mixer is 1 m^3 per hour

Therefore cost of mixing per m^3 = $\frac{£5.49}{1\ m^3}$ = £5.49

Therefore cost of mixed mortar = £51.27 + £5.49
= £56.76 per m^3

Since 0.05 m^3 of mortar is required per square metre of brickwork, one brick thick, the cost per m^2 = £56.76 × 0.05
= £2.83 . . . [2]

(b) *Labour*:

Assume 2 + 1 gang
Gang cost per hour = £(4.50 + 4.50 + 4.00) = £13.00
Assume 1 bricklayer can lay 60 bricks per hour

Therefore cost per m^2 = $\frac{£13.00 \times 118}{(60 + 60)}$ = £12.78 . . . [3]

Total rate ([1] + [2] + [3]) = <u>£28.35 per m^2</u>

Item B – Facings half-brick thick in skin of hollow wall in cement-lime mortar (1 : 2 : 9) pointed with a neat flush joint as work proceeds

Basic prices:

Facing bricks £250.00 per thousand + £12.00 crane off-loading
Cement and sand as before
Lime £70.00 per tonne

Method

(a) *Materials*:

Cost of facing bricks delivered to site per 1,000	= £250.00
Add crane off-loading	= 12.00
Add 1 hr per 1,000 assistance with crane off-loading	= 4.00
	£266.00
Add 12 % waste	31.92
	£297.92

Number of bricks required per m^2 = 59

Therefore cost per $m^2 = \dfrac{£297.92 \times 59}{1{,}000} = £17.58$ . . . [1]

Mortar – Cost of lime delivered to site per tonne	= £ 70.00
Allow 1 hr unloading and stacking	= 4.00
	£ 74.00

Cost of (1 : 2 : 9)mix:

1 m^3 of cement costs (from before)	= £103.68
2 m^3 of lime cost £74.00 × 0.60 × 2	= 88.80
9 m^3 of sand cost £11.00 × 1.52 × 9	= 150.48
	£342.96
Add 33⅓% for shrinkage and waste	114.32
Cost per 12 m^3	= £457.28

Therefore cost per $m^3 = \dfrac{£457.28}{12} = £38.11$

Mixing cost as before, i.e. £5.49, therefore total cost of mortar = £38.11 + £5.49 = £43.60 per m^3

Since 0.018 m^3 of mortar is required per square metre of brickwork half-brick thick, the cost per m^2 = £43.60 × 0.018
= 79p . . . [2]

(b) *Labour*:

Assume 2 + 1 gang – gang cost per hour as before £13.00
Assume 1 bricklayer can lay 40 bricks per hour including pointing up as work proceeds

Therefore cost per $m^2 = \dfrac{£13.00 \times 59}{(40 + 40)} = £9.59$ . . . [3]

Total rate ([1] + [2] + [3]) = <u>£27.96 per m^2</u>

Item C – Extra over common brickwork for facings (PC £200.00 per 1,000 delivered to site) in cement-lime mortar (1 : 2 : 9) in English bond pointed with a neat flush joint as work proceeds

Basic prices:

As before.

Method

(a) *Extra cost of materials*:

Number of facing bricks required per m² of wall in English bond = 89

Cost of commons delivered to site per 1,000	=	£100.00
Add 8% waste	=	8.00
		£108.00

Therefore cost of 89 commons = $\frac{£108.00 \times 89}{1,000}$ = £9.61

Cost of facings delivered to site per 1,000	=	£200.00
Allow 1½ hrs per 1,000 unloading at £4.00	=	6.00
		£206.00
Add 12% waste	=	24.72
		£230.72

Therefore cost of 89 facings = $\frac{£230.72 \times 89}{1,000}$ = £20.53

Therefore extra over cost of bricks per m²
= £20.53 – £9.61 = £10.92 . . . [1]

(*Note*: No adjustment needs to be made for thc cost of mortar since the full cost will have been taken into account in the pricing of the common brickwork item.)

(b) *Extra cost of labour*:

Cost of laying 89 commons (from previous example) =
$\frac{£13.00 \times 89}{(60 + 60)}$ = £9.64

Cost of laying 89 facings (from previous example) =
$\frac{£13.00 \times 89}{(40 + 40)}$ = £14.46

Therefore extra over cost of labour per m²
= £14.46 – £9.64 = £4.82 . . . [2]

Total rate ([1] + [2]) = <u>£15.74 per m²</u>

Item D – Form cavity 50 mm wide in hollow wall with 4no. wall ties per m²

Basic prices:

Galvanised butterfly pattern wall ties £4.50 per 100

Method

(a) *Materials*:

Cost of wall ties delivered to site per 100 = £4.50
Add 5 % waste 0.23
£4.73

$$\text{Therefore cost per m}^2 = \frac{£4.73 \times 4}{100} = 19\text{p} \quad \ldots [1]$$

(b) *Labour*:

Assume 2 + 1 gang at £13.00 per hour can form 25 m^2 of cavity per hour

$$\text{Therefore cost per m}^2 = \frac{£13.00}{25} = 52\text{p} \ldots [2]$$

Total rate ([1] + [2]) = <u>71p per m^2</u>

Item E – Close 50 mm cavity at jambs with common brickwork

Basic prices:

As before.

Method

(a) *Materials*:

Bricks – Allow approx. 7no. bricks at 11p each allowing for waste (see previous example) = 77p . . . [1]
Mortar – Allow a nominal cost of, say 10p . . . [2]

(b) *Labour*:

Assume 2 + 1 gang can close 10 m of cavity per hour

$$\text{Therefore cost per m} = \frac{£13.00}{10} = £1.30 \quad \ldots [3]$$

Total rate ([1] + [2] + [3]) = <u>£2.17 per m</u>

Item F – 112 mm wide d.p.c. laid horizontally

Basic prices:

112 mm wide polythene d.p.c. per 30 m roll £5.50

Method

(a) *Materials*:

Cost of 1 roll of d.p.c. delivered to site = £5.50
Add 5% waste and laps = 0.28
£5.78

$$\text{Therefore cost per m} = \frac{£5.78}{30} = 19\text{p} \quad \ldots [1]$$

(b) *Labour*:
Assume 2 + 1 gang can lay and bed 1 roll of d.p.c. in 0.12 hr

$$\text{Therefore cost per m} = \frac{£13.00 \times 0.12}{30}$$
$$= 5\text{p} \qquad \ldots [2]$$

Total rate ([1] + [2]) = <u>24p per m</u>

Item G – 100 mm thick block wall in cement-lime mortar (1 : 2 : 9) finished fair one side

Basic prices:
440 × 215 × 100 mm thick blocks £12.00 per m²

Method

(a) *Materials*:

Blocks – Cost of blocks delivered to site per m²	= £12.00	
Allow 0.1 hr per m² assistance with unloading	= 0.40	
	£12.40	
Add 5%waste	0.62	
	£13.02	. . . [1]

Mortar – Cost of mixed mortar per m³ from previous example £43.60
Therefore cost of mortar per m² = £43.60 × 0.0066 m³
= 29p . . . [2]

(b) *Labour*:
Assume 2 + 1 gang can lay 4 m² in 1 hr

$$\text{Therefore cost per m}^2 = \frac{£13.00}{4} = £3.25$$

Add for fair face and pointing at 10 m² per hr for gang

$$\text{Therefore cost per m}^2 = \frac{£13.00}{10} = £1.30$$

Therefore total labour cost = £3.25 + £1.30 = £4.55 . . . [3]

Total rate ([1] + [2] + [3]) = <u>£17.86 per m²</u>

Item H – Hole for small pipe through half brick thick wall in facings and make good one side

Method

(a) *Materials*:

Bricks – Allow approx. 2no. bricks at 23p each allowing for waste (see previous example) = 46p . . . [1]

Mortar – Allow a nominal cost of, say 5p . . . [2]

(b) *Labour*:

Assume 1 bricklayer can cut hole and make good in 0.3 hr

Therefore cost = £4.50 × 0.3 = £1.35

Allow for 50% labourer in attendance

= £4.00 × 0.15 = 0.60 . . . [3]

£1.95

Total rate ([1] + [2] + [3]) = £2.46 each

7.5 Roofing

There are many different types of materials available for both flat and pitched roofs of various construction, the more common forms of covering being provided for in the Standard Method under the following headings:

1 Slating and tiling.
2 Sheet roofing and cladding in various profiles.
3 Roof decking, e.g. wood wool slabs, etc.
4 Built-up bitumen felt roofing.
5 Sheet metal roofing, flashings and gutters.

In addition to the above, mastic asphalt is widely used as a covering to flat roofs, but is measured under section L – 'Asphalt work'.

7.5.1 Slating and tiling

Slating may comprise either natural or man-made slates in varying sizes ranging from 250 × 150 mm to 650 × 400 mm. Natural slates are obtained principally from parts of Wales and Cumbria, the latter being considered generally to be of better quality, reflected in the higher price. Alternatively, cement fibre or similar artificial slates are manufactured which are much cheaper than their naturally occurring counterparts.

Slates are fixed with copper or alloy nails to softwood roofing battens with two nail holes being punched in the centre or near to the head of each slate depending on how they are to be fixed.

Tiles may be manufactured from natural clay or concrete in various styles, colours and sizes ranging from 265 × 165 mm, the standard size for plain tiles, to 420 × 330 mm for the various types of interlocking tiles.

Tiles are supplied with one or two holes situated near the head for securing with nails, as above; however, in addition each tile has two nibs projecting on the underside again, near the head for hanging from battens. By making this secondary provision for fixing, it means that not every tile needs to be nailed but only, say, one course in four. Slates, however, having no nibs, must be individually nailed.

Accessories

Special matching hip, valley and ridge tiles in various materials are manufactured for use in conjunction with both slating and tiling.

Extra wide tiles/slates known as 'tile and a half' or 'slate and a half' are used at verges in alternate courses to break the joint in plain tiling and plain slating work.

Method of pricing

(a) Materials. The Standard Method requires the following items to be included in the bill description, each of which must be priced and expressed in terms of cost per square metre.

1 The slates/tiles.
2 The battens on to which the slates/tiles are fixed.
3 The nails needed to fix both the slates/tiles and the battens.

Slates and tiles are purchased by the thousand which means that the number of slates/tiles needed to cover 1 m^2 of roof must be calculated in the first instance.

The total area measured in the bill corresponds to the **net** area of roof covering, i.e. no allowance being made for laps, and since in plain work a substantial part of the slate/tile is lapped under the tile above and over the tile below, the **net** area of roof covered by each tile is very small.

Example

Assume 265 × 165 mm plain tiles laid to a lap of 63 mm.

Net area of roof covered by 1 tile

$$= \frac{\text{gauge (i.e. length} - \text{lap)} \times \text{width}}{2}$$

$$= \frac{(265 - 63\text{ mm})}{2} \times 165\text{ mm}$$

$$= 101 \times 165\text{ mm} = 0.0167\text{ m}^2$$

Therefore number of tiles required to cover 1 m^2 of roof

$$= \frac{1}{0.0167} = 60$$

In order to illustrate the extent of the area which is 'lost' due to the laps, if these 60 tiles were placed end to end they would cover an area of:

$$60 \times 0.265 \times 0.165\text{ m} = 2.62\text{ m}^2$$

This allowance for laps must be made by the estimator in his price and by the firm's buyer when placing an order.

Where interlocking tiles are used, this loss of area is greatly reduced since both long edges of the tiles are watertight due to the interlocking design and the need for tiles to be positioned immediately below the vertical joints is therefore eliminated. Thus, fewer are needed per square metre of roof as shown below.

Example
Assume 420 × 330 mm interlocking tiles laid to a 75 mm lap

Net area of roof covered by 1 tile
= (length − head lap) × (width − side lap)
= (420 − 75 mm) × (330 − 25 mm)
= 345 × 305 mm
= 0.1052 m²

Therefore number of tiles needed to cover 1 m² of roof

$$= \frac{1}{0.1052} = 9.51$$

It can now be readily appreciated why interlocking tiling is much cheaper than plain tiling.

The battens are purchased by the metre or 100 m and therefore the total length in metres per square metre of roof must be calculated.

Since each tile/slate must be affixed in some way to a batten, the number of each type of tiles/slates required per square metre multiplied by the **net** width of the tile/slate will give the total length of battens needed.

Example:
For plain tiles 165 mm wide – no side laps, effective width = 165 mm
Total length of battens required per m² = 60 × 165 mm
= 9.90 m

For interlocking tiles 330 mm wide, effective width
= 330 mm − side lap
= 330 − 25 mm = 305 mm
Total length of battens required per m² = 9.51 × 305 mm
= 2.90 m

Nails are purchased by weight with 40 mm alloy nails weighing approximately 3 kg per 1,000. The number of nails per square metre of roof can be calculated by adding together those quantities required for the tiles/slates and those for nailing the battens to the rafters.

(b) Labour. For large areas of roofing, it is normal practice to organise the labour into gangs comprising usually of 2 tilers who will fix the battens and tiles, and 1 labourer whose job it will be to keep the craftsmen supplied with materials.

7.5.2 Sheet roofing and cladding

Sheet materials provide a quick, durable and relatively cheap method of covering to sloping and vertical surfaces with accessories available for fixing to timber or steel backings. The more common types of materials used include corrugated fire resistant cement sheeting, galvanised corrugated iron sheeting, corrugated translucent sheeting and pvc coated galvanised steel or aluminium sheeting in a range of profiles and colours.

Method of pricing

(a) Materials. The bill rate per square metre must include the following materials:

1 The sheeting material.
2 The hook bolts and washers (for fixing to steel backgrounds) or drive screws and washers (for fixing to timber backings).

As with tiled roofing, the measured area in the bill makes no allowance for laps, and since all types of sheeting are usually lapped 150 mm at the ends of the sheet and approximately 75 mm at the sides, again the **net** area of roof covered by each sheet must be calculated. Where a range of sheet sizes is available, the percentage allowance for laps will vary as illustrated below.

Example:

For 'Big six' corrugated fire resistant cement sheeting in sheet sizes 1,086 mm wide × 2,120 mm long:

Effective width = 1,086 − 75 mm = 1,011 mm
Effective length = 2,120 − 150 mm = 1,970 mm
Gross area of sheet = 1.086 × 2.120 = 2.302 m^2
Net area of sheet = 1.011 × 1.970 = 1.992 m^2
Area of laps = 2.302 − 1.992 m^2 = 0.31 m^2

$$\text{Allowance for laps} = \frac{0.31}{1.992} \times 100\% = 15.56\%$$

The method of fixing sheet materials would be in accordance with the manufacturer's instructions, but a typical 2 × 1 m sheet may be secured by six screws or hook bolts, two at the head, two at the tail and two at an intermediate fixing point.

7.5.3 Roof decking

Roof decking comprises flat slabs in wood wool, strawboard or other similar man-made material with various types of finish applied such as prefelted, prescreeded or with the soffit pretextured. The slabs may be obtained plain or with steel channel reinforced edges for spanning greater distances. The thickness of the slabs ranges from 50 mm to 125 mm in 25 mm stages, and whilst the width of the slabs is a standard 600 mm, the length can vary from 1,800 mm to 4,000 mm.

Two methods of fixing the slabs include securing to timber joists with large-headed galvanised nails at 150 mm centres at each support, and to steel or concrete beams with proprietary fixing clips at similar distances.

Both skilled and unskilled labour would normally be employed on fixing the slabs, possibly in the ratio of 1 : 1 depending on the area, height above ground, etc.

7.5.4 Built-up felt roofing

Built-up felt roofing is invariably carried out by specialist firms and is not covered here.

7.5.5 Sheet metal roofing, flashings and gutters

Sheet metal roofing such as lead, copper, aluminium, etc. is comparatively rare today due to its very high cost, but where this type of roofing is specified, again, it would be carried out by specialists. However, lead or lead-based flashings are widely used in normal building work and form an important part of most roofing systems.

(a) Materials

Lead for flashings is supplied in rolls of various thicknesses ranging from 1.25 mm and weighing 14.18 kg/m^2 (colour-coded green for identification on site) to 3.55 mm and weighing 40.26 kg/m^2 (colour-coded orange).

The rolls are a standard 6 m long and are available in eight different widths: 150, 210, 240, 270, 300, 390, 450 and 600 mm.

(b) Method of fixing

Where large quantities of flashings are required, the fixing operation is often carried out by a gang comprising a plumber and mate. Rolls of suitable width are selected which must then be cut to size and fixed with the top edge of the lead turned into a groove or joint in the brickwork and secured by lead wedges prior to pointing in mortar, whilst the bottom or free edge should be retained with lead or copper clips at the specified centres.

(c) Method of pricing

Flashings are measured in metres stating the width, lap and method of fixing with the **net** length being taken making no allowance for laps or seams in the running length. The estimator must therefore make the necessary percentage allowance for this 'loss' of material in his price.

Example:
Consider a length of lead flashing where 100 mm laps are required at centres not exceeding 2 m
Net length of each piece of lead = 2,000 − 100 mm lap
= 1,900 mm

Therefore allowance for laps = $\frac{0.10}{1.90} \times 100\% = 5.26\%$

It can be assumed that lead clips and wedges would be made from offcuts and as such included in the normal waste allowance.

Roofing – guide to labour outputs

Slating (based on 2 + 1 gang)		Number fixed per hour by gang
Slate size	350 × 200 mm	80–100
	400 × 250 mm	65–70
	600 × 350 mm	45–50
Tiling (based on 1 + 1 gang)		
Plain tile size	265 × 165 mm	110–130
Interlocking the size	420 × 330 mm	40–60

Battens: Approx. 30–40 m per hour (craftsman only), including setting out. The output of tiling is higher than that for slating as the holes are already punched in the tile and not every tile requires nailing. This also explains the difference in the gang strength.

Sundry labours:
(a) Raking cutting on slating: 5–6 m per hour (based on 2 + 1 gang) + waste on slates.
(b) Double course of tiles at eaves: 20 m per hour (based on 1 + 1 gang).
(c) Raking cutting on plain tiling: 2 m per hour.
(d) Raking cutting on interlocking tiling: 3 m per hour.

Sheeting (based on 1 + 1 gang)
To sloping surfaces: Allow 10 min per m^2 of sheet, with an addition of 10 per cent where fixed with hook bolts in lieu of drive screws. For vertical cladding, allow an extra 25 per cent.

Roof decking (based on 1 + 1 gang)
Allow between 5 and 10 m^2 per hour depending on thickness and type of slab.

Lead flashings (based on 1 + 1 gang)
(a) Cutting and fixing 225 mm girth lead flashing in metres/hour.
(b) Flashings and aprons: 3 m.
(c) Stepped flashings: 1.5 m.

The typical bill items to be priced in this section are as follows:

					£	p
A	450 × 225 natural Welsh slating fixed with 32 copper nails and laid to a 75 lap on and including 38 × 19 battens	425	m^2			
B	Raking cutting	75	m			
C	265 × 165 plain concrete tiles fixed with 32 alloy nails every fourth course and laid to a 63 lap on and including 38 × 19 battens	670	m^2			
D	420 × 330 concrete interlocking tiles fixed with 32 nails every fourth course and laid to a 75 lap on and including 38 × 19 battens	1,020	m^2			
E	Extra for half-round ridge tiles bedded and pointed in cement mortar (1 : 3)	112	m			
F	Corrugated fire resistant cement sheeting lapped one and a half corrugations at sides and 150 at ends, fixed with hook bolts and washers to steel purlins	590	m^2			
G	50 thick wood wool slabs fixed with galvanised nails to softwood bearers	1,120	m^2			
H	Code 3 lead flashing weighing 14.18 kg/m^2 150 girth fixed with lead tacks and laid with 100 laps, top edge wedged into groove with lead wedges and pointed in cement mortar (1 : 3) (measured net)	84	m			

The build-up of roofing rates

Item A – 450 × 225 roof slating

Basic prices:

All-in rate for craftsman: £4.50 per hr
All-in rate for labourer £4.00 per hr
450 × 225 natural Welsh slates £1,400.00 per 1,000
32 copper nails £5.00 per kg
38 × 19 sawn softwood battens £24.00 per 100 m

Method

(a) Materials:

Cost of slates delivered to site per 1,000 = £1,400.00
Allow $1\frac{1}{2}$ hrs per 1,000 unload and stack = 6.00

£1,406.00

Number of slates required per m² = gauge × width

$$\text{gauge} = \frac{\text{length} - \text{lap}}{2} = \frac{450 - 75}{2} = 188 \text{ mm}$$

Therefore effective area of 1 slate = 0.188 × 0.225
= 0.0423 m²

Therefore number required per m² of roof

$$= \frac{1}{0.0423} = 24$$

$$\text{Therefore cost per m}^2 = \frac{£1{,}406.00 \times 24}{1{,}000} = £33.74$$

Nails – Each slate is nailed with 2 copper nails. Therefore number of nails per m²
= 2 × 24 = 48
There are approx. 360 copper nails per kg

$$\text{Therefore cost of nails per m}^2 = \frac{£5.00 \times 48}{360} = 67\text{p}$$

Battens – Cost of battens delivered to site per 100 m = £24.00
Unload and stack $\frac{1}{4}$ hour per 100 m 1.00

£25.00

Quantity required per m² = 24 × 0.225 = 5.4 m

$$\text{or} \quad \left(\frac{1{,}000}{\text{gauge}} = \frac{1{,}000}{188} = 5.32 \text{ m}\right)$$

$$\text{Therefore cost per m}^2 \text{ of roof} = \frac{£25.00 \times 5.4}{100} = £1.35$$

Allow for nails to battens (say) 0.20

£1.55

Total material cost per m² = £35.96
Add 5% waste 1.80

£37.76 . . .[1]

(b) *Labour*

Fixing slates –
Assume 2 + 1 gang. Therefore hourly gang cost = 2 × £4.50 + £4.00 = £13
Assume no. of slates fixed by gang in 1 hr = 70

$$\text{Therefore cost per m}^2 = \frac{£13.00 \times 24}{70} = £4.46$$

Fixing battens – Assume 1 slater can set out and fix 35 m of battens per hour

Therefore cost per m² = $\frac{£4.50 \times 5.4}{35}$ = 69p

Total labour cost per m² = £5.15 . . .[2]

Total rate ([1] + [2]) = £42.91 per m²

Item B – Raking cutting

Assumptions

The number of slates which would require cutting per m would be equal to 1 m/margin, the margin being the length of the exposed portion of the slate on the roof. Since the margin is equal to the gauge, the number of slates involved would be 1m/188 = 5.4no. If the length of the cut on each slate is, say, 400 mm the total length of cutting would be 5.4 × 0.40 = 2.16 m, which means that for every 1 m of raking cutting measured in the bill of quantities, 2.16 m of actual cutting on site needs to be carried out.

Method

(a) *Materials*:

Assume 50 per cent of the 5.4 slates which are cut is wasted. Therefore, 2.7 (say 3) slates should be allowed for
3 slates at £1,406.00 per 1,000 = £4.22 . . . [1]

(b) *Labour*

Based on gang cost of £13.00 per hour as before
Assume gang can cut 6 m of slates per hour (as measured in the bill, i.e. equivalent to 2.16 × 6 m = 12.96 m of actual cutting)

Therefore labour cost = $\frac{£13.00}{6}$ = £2.17 . . . [2]

Total rate ([1] + [2]) = £6.39 per m

Item C – 265 × 165 plain concrete tiling

Basic prices:

265 × 165 plain concrete tiles £300.00 per 1,000
Softwood battens £24.00 per 100 m
32 mm alloy nails £5.50 per kg

Method

(a) Materials:

Cost of 265 × 165 plain concrete tiles delivered to site per 1,000 = £300.00
Allow $1\frac{1}{2}$ hrs per 1,000 unload and stack = 6.00

£306.00

No. of tiles required per m² = gauge × width

$$\text{gauge} = \frac{265 - 63}{2} = 101 \text{ mm.}$$

Therefore effective area of 1 tile
= 0.101 × 0.165
= 0.01667 m². Therefore number required per m² of roof

$$= \frac{1}{0.01667} = 60$$

$$\text{Therefore cost per m}^2 = \frac{£306.00 \times 60}{1{,}000} = £18.36$$

Nails – Tiles are nailed every fourth course, therefore number of tiles nailed in 1 m² are 60/4 = 15 × 2 nails per tile = 30 nails
There are approx. 350 alloy nails per kg

$$\text{Therefore cost per m}^2 = \frac{£5.50 \times 30}{350} = 47\text{p}$$

Battens – Cost of battens per 100 m (from before) = £25.00
Quantity required per m² of roof = 60 × 0.165
= 9.9 m
(or 1000/gauge = 1000/101)

Therefore cost per m² of roof

$$= \frac{£25.00 \times 9.9}{100} = £2.48$$

Allow for nails to battens, say = 0.35

£2.83

Total material cost per m² = £21.66
Add 5% waste = £ 1.09

£22.75 . . . [1]

(b) *Labour*:

Based on 1 : 1 gang cost of £4.50 + £4.00 = £8.50 per hour
Assume number of tiles fixed by gang in 1 hr = 120

$$\text{Therefore cost per m}^2 = \frac{£8.50 \times 60}{120} = £4.25$$

Assume 1 tiler can set out and fix 35 m of battens per hour

$$\text{Therefore cost per m}^2 = \frac{£4.50 \times 9.9}{35} = £1.27$$

Therefore total labour cost = £5.52 . . . [2]

Total rate ([1] + [2]) = <u>£28.27 per m²</u>

Item D – 420 × 330 concrete interlocking tiles

Basic prices:

420 × 330 concrete interlocking tiles £800.00 per 1,000
Softwood battens £24.00 per 100 m
32 mm alloy nails £5.50 per kg

Method

(a) *Materials*

Cost of 420 × 330 concrete interlocking tiles delivered to site per 1,000 = £800.00
Allow 2 hrs per 1,000 unload and stack = 8.00
£808.00

Number of tiles required per m^2 = (length – lap) × (width – side lap)
= (420 – 75) × (330 – 25)
= 0.1052 m^2

Therefore number required per m^2 of roof = 1/0.1052 = 9.51

Therefore cost per m^2 = $\frac{£808.00 \times 9.51}{1,000}$ = £7.68

Nails – Tiles are nailed every fourth course; therefore number of tiles nailed in 1 m^2 = 9.51/4 = 2. Cost is therefore negligible – allow a nominal 10p

Battens – Cost of battens per 100 m (from before) = £25.00
Quantity required per m^2 of roof = 0.305 × 9.51
= 2.9 m

Therefore cost per m^2 of roof = $\frac{£25.00 \times 2.9 \text{ m}}{100}$ = 73p

Allow for nails to battens, say 5p
78p

Total material cost = £8.46
Add 5% waste 0.43
£8.89 . . . [1]

(b) *Labour*:

Based on 1 : 1 gang cost of £8.50 per hour
Assume number of tiles fixed by gang in 1 hr = 50

Therefore cost per m^2 = $\frac{£8.50 \times 9.51}{50}$ = £1.62

Assume 1 tiler can set out and fix 35 m of battens per hour

Therefore cost per m^2 = $\frac{£4.50 \times 2.9 \text{ m}}{35}$ = 37p

Total labour cost = £1.99 . . . [2]

Total rate ([1] + [2]) = <u>£10.88 per m^2</u>

Item E – Half-round ridge tiling

Basic prices:

Ridge tiles 456 mm long £250.00 per 100

Method

(a) *Materials*:

Cost of ridge tiles delivered to site per 100		= £250.00
Allow ½ hr per 100 unload and stack		= 2.00
		£252.00
Number of tiles required per m = 1,000/456 =2.19		
Therefore cost per m = $\frac{£252.00 \times 2.19}{100}$	= £5.52	
Allow 5% waste	0.28	
	£5.80	
Allow for mortar per m, say	0.50	
	£6.30	. . . [1]

(b) *Labour*:

Based on 1 : 1 gang cost of £8.50 per hour

Assume gang can fix, bed and point 15 ridge tiles per hour

Therefore cost per m = $\frac{£8.50 \times 2.19}{15}$ = £1.24 . . . [2]

Total rate ([1] + [2]) = <u>£7.54 per m</u>

Item F – Fire resistant sheet material

Basic prices:

1,825 × 2,120 mm fire resistant cement sheeting £15.00 per sheet
Galvanised hook bolts and nuts £10.50 per 100
Plastic washers £3.50 per 100

Method

(a) *Materials*:

Cost of fire resistant cement sheeting delivered to site	= £15.00 each
Unload and stack 1/10 hr per sheet	= 0.40
	£15.40

Hook bolts and nuts – Assume 3 rows of fixings at 3 bolts per row = 9 bolts

Therefore cost of bolts and nuts per sheet

= $\frac{(£10.50 + £3.50) \times 9}{100}$ = £1.26

Material cost per sheet =	£16.66
Allow 5% waste	0.83
	£17.49

Effective area of sheet allowing for laps
= (1,825 − 75) × (2,120 − 150)
= 1,750 × 1970
= 3.45 m^2

Therefore material cost per m^2 = $\frac{£17.49}{3.45}$ = £5.07 . . . [1]

(b) *Labour*:
Based on 1 : 1 gang cost of £8.50 per hour
Assume gang can fix 1 m^2 of sheeting in 1/6 hr
Therefore cost per m^2 = $\frac{£8.50}{6}$ = £1.42
Add 10% for fixing with hook bolts in lieu of
drive screws = 0.14
£1.56 . . . [2]

Total rate ([1] + [2] = £6.63 per m^2

Item G – 50 mm thick wood wool slabs

Basic prices:
600 × 2,000 × 50 mm thick wood wool
Slabs £3.40 per m^2
100 mm galvanised wire slab nails £16.50 per 10 kg

Method

(a) *Materials*:
Cost of 50 mm thick wood wool slabs delivered to
site per m^2 = £3.40
Unload and stack 1/20 hr per slab (1.2 m^2) = 0.17
£3.57

Number of nails required:
Assume 3 rows of fixings with nails at 150 mm centres
Therefore number required per slab = $\frac{(3 \times 600)}{150}$ = 12

There are approx. 100 slab nails per kg
Therefore cost of nails per slab = $\frac{£16.50 \times 12}{10 \times 100}$ = 0.20

Therefore cost per m^2 = $\frac{0.20}{0.60 \times 2.00}$ = 17p

Total material cost per m^2 = £3.74
Allow 5% waste = 0.19
£3.93 . . . [1]

(b) *Labour*:

Based on 1 : 1 gang cost of £8.50 per hour

Assume gang can lay and fix 10 m^2 of slab per hour

Therefore cost per m^2 = $\frac{£8.50}{10}$ = 85p . . . [2]

Total rate ([1] + [2]) = <u>£4.78 per m^2</u>

Item 4 – Code 3 lead flashing 150 mm girth

Basic prices:

Code 3 lead flashing weighing 14.18 kg/m £850.00 per tonne

Method:

(a) *Materials*:

Cost of Code 3 lead delivered to site per tonne	= £850.00
Allow 2 hrs per tonne unload and stack	= 8.00
	£858.00
Allow 7½% for waste and laps	£64.35
	£922.35

Total area per tonne of Code 3 lead = $\frac{1{,}000 \text{ kg}}{14.18}$ = 70.52 m^2

Therefore total length of flashing 150 mm wide per tonne

= $\frac{70.52}{0.150}$ = 470.13 m

Therefore cost per m = $\frac{£922.35}{470.13}$ = £1.96

Assume off-cuts of lead are used for wedges and tacks

Allow mortar for pointing, say = 0.05

Total material cost = £2.01 . . . [1]

(b) *Labour*:

Based on gang cost of plumber and mate at

£6.00 + £4.00 = £10.00 per hour

Assume gang can cut and fix 5 m of flashing per hour

Therefore cost per m = $\frac{£10.00}{5}$ = £2.00 . . . [2]

Total rate ([1] + [2]) = <u>£4.01 per m</u>

7.6 Woodwork

In recent years, the trend towards the mass production of standard units by a relatively small number of large manufacturers has reduced much of the former 'carpentry and joinery' work to a matter of fixing or assembling on site factory-made components such as roof trusses, door sets, windows, kitchen units, etc. However, skilled labour is still required for most site operations involving the use of wood or wood-based products.

The Standard Method arranges the work in the woodwork section into the following six main categories which generally follow the sequence in which the work would be carried out on site:

1. *Carcassing*. Involving structural timbers forming the framework of roofs, floors, walls, etc.
2. *First fixings*. Items generally which are fixed after the framework and including roof and floor boarding together with grounds, battens and bearers, etc. which form a base or backing to which other work is subsequently fixed.
3. *Second fixings*. Covering: (1) Those items of 'dressed' or planed wood, many of which are obtainable in standard sizes 'off the shelf' from timber merchants and which are exposed in the finished work and require decoration. For example, skirtings, architraves, window boards, etc. (2) Sheet materials fixed as wall and ceiling linings having a wood 'base' such as plywood, plastic-covered chipboard and other proprietary boards.
4. *Composite items*. Factory-made standard items supplied ready for assembling and fixing on site.
5. *Sundries*. General labours in connection with woodworking operations, together with the supply of insulating materials and metalwork.
6. *Ironmongery*. Involving all types of door, window ironmongery, etc.

7.6.1 Materials

Softwoods

Carcassing timbers, first fixings and the bulk of 'joinery' quality timber used in building work in this country consist of softwoods of one type of another, those in most common use being Redwoods, Whitewoods and Hemlock which are obtained principally from Northern Russia, Finland, Scandinavia and North Eastern Europe. The specification in the trade preamble will identify the grade or quality required enabling the estimator to price accordingly, the cost of timber varying considerably depending on the quality.

With the exception of boarding and flooring, most carcassing and first-fix items are supplied sawn, whilst timber for second-fix and composite items are converted from larger sections into standard widths and thicknesses and planed either on one side and one edge (PSE) or all round (PAR) and is of better quality, the best originating from the area around the Kara Sea in Northern Russia and shipped from the port of Archangel.

Hardwoods

A wide range of hardwoods, some being suited to specific purposes, is available where high-class work is required and the woodwork is to be a main feature of the building. As with softwoods, the price of hardwood varies according to the type chosen and therefore the particular variety of wood must be specified in the bill. Any work in hardwood tends to be very expensive for a number of reasons:

1 It has a high basic cost since many hardwoods come from tropical parts of the world and handling and shipping costs are high.
2 In order to prepare the timber for sale, a great deal of work has to be carried out, including proper seasoning, kiln drying often for long periods, machining, etc.
3 Due to its denseness, pattern of grain and sometimes coarseness, the length of time needed to work on the hardwood and fix it in position is invariably much greater than that for a corresponding item in softwood. Teak, for example, has a grain which contains minute particles of silica making it difficult to work and blunting tools very quickly.
4 Extra care must be taken when selecting lengths of hardwood where the grain and colour is required to match up.
5 A higher waste allowance compared with softwood must be made due to the selection process mentioned above and a higher cutting waste because of a greater degree of unevenness in the wood. Hardwoods, in most cases, are chosen for clear finish and those lengths of poor appearance may have to be discarded.
6 Hardwoods are often secured in position using relatively expensive methods of fixing such as screws, brass cups and screws, pelleting, etc.

Depending on the type of hardwood used, the extra labour required would vary between 50 and 150 per cent more than that needed for a comparable item in softwood.

Method of pricing

The normal unit of purchase of carcassing timber is the cubic metre, whilst the unit of measurement is the metre; however, most items of manufactured joinery are sold by the metre and no further conversion is needed.

The price of timber is sometimes quoted 'FOT', meaning 'free on transport'. In other words, the price includes only for unloading the timber from the ship and on to the buyer's own transport, the estimator then having to allow in addition for the cost of transport and delivery to the site or workshop. Similarly, prices quoted 'FOB' and 'FOQ' mean 'free on board' and 'free on quay' respectively, and again the buyer must allow for unloading/loading in addition to transport costs as before.

Means of fixing

Items of woodwork are deemed to be fixed by nails unless otherwise stated, which makes it necessary for the estimator to calculate the cost of those required per cubic, linear or square metre of the item concerned. For example:

General carcassing timbers – 2 kg per cubic metre
Boarding and flooring – 5 nails per metre

Where items are plugged to the background, the distance between the centres of the plugs will be stated, allowing the estimator to calculate the number per metre required together with the quantity of nails assuming two nails per plug.

Other methods of fixing include bolting, shot firing (to steelwork), screwing or even glueing, the particular method being stated in the bill description. For items which are bolted, the bolts themselves and forming the hole are measured separately; for items which are fixed by screws, allow 25 per cent additional labour, with screwing and pelleting requiring approximately 50 per cent extra labour.

Other extra costs

In addition to the 'basic' price per cubic metre quoted by a supplier, the following additional costs may also be incurred where the bill of quantities specifies:

1 *Timber impregnation with preservative*. Where timber is exposed to conditions which might encourage decay, the wood can be impregnated under pressure with special preservative solutions which help to protect it. The cost of this treatment would be quoted in terms per cubic metre and may be in the region of 10 per cent of the basic price.
2 *Long lengths*. Where single lengths of timber are required over 4.8 m long, the cost usually increases on a sliding scale as follows:

Lengths:	up to 4.8 m	– basic price
	exc. 4.8 m n.e. 5.4 m	– basic price + 7½%
	exc. 5.4 m n.e. 6.0 m	– basic price + 15%, etc.

3 *Stress grading*. Structural timbers are required to be stress graded to a standard specified in the bill at the timber merchant's premises where the wood is passed through a machine and stamped with a colour code to signify its strength. The cost of providing this service is approximately 10 per cent of the basic cost.

4 *Small sections and angle fillets*. Basic prices of timber normally apply to square or rectangular sections over 75 × 50 mm. Where smaller sections or angle and tilting fillets are required, additional sawing is needed, the cost of which is added to the price of the timber and may be in the order of 7½–10 per cent of the basic cost.

Waste

Waste is inevitable with all timber and sheet materials, the extent varying according to the nature of the material and the operation being carried out and whilst 10 per cent allowance may be sufficient for carcassing items, anything between 5 per cent and 25 per cent if not more may be needed for some second-fix items. Where items of manufactured joinery such as doors and windows are concerned, a small allowance needs to be made to cover the occasional broken or damaged item – say 2½ per cent. Whilst there is no cutting waste involved on ironmongery, an allowance of approximately 2½ per cent should be made in anticipation of losses through theft and breakages.

Ironmongery

Ironmongery is normally supplied with its own matching screws and so no further fixing materials need to be allowed for. The length of time taken to fix an item of ironmongery depends on the nature of the background – i.e. softwood, hardwood, brick, etc. – the location of the item – i.e. degree of accessibility – and the preparatory work required prior to fixing – i.e. cutting mortices, sinkings, boring for holes, etc.

7.6.2 Woodwork – Guide to labour outputs

Outputs apply to skilled operatives:

Structural members of roofs and floors (wall plates, rafters, ceiling joists, etc.)	10 m per hour
Partitions	6 m per hour
Fascia and barge boards	8 m per hour
Grounds generally	12 m per hour
Herringbone strutting	5 m per hour
Door frame/lining for single door	0.75 hr each
Tongued and grooved flooring	12 m per hour
Chipboard flooring	10 m^2 per hour
Roof trusses (based on 1 craftsman + 3 labourers)	3 no per hour
Hollow-cored single door	0.75 hr each
100 mm skirting, including mitres	12 m per hour

The typical bill items to be priced in this section are as follows:

					£	p
A	150 × 50 roof joists	560	m			
B	50 × 50 herringbone strutting to 175 joists	255	m			
C	150 × 25 softwood skirting on and including 25 × 12 grounds plugged to brickwork at 300 centres	782	m			
D	1,225 × 1,073 softwood window	60	no.			
E	18 thick external quality plywood flat roof boarding	900	m^2			
F	25 thick softwood tongued and grooved board flooring	512	m^2			
G	838 × 1,981 × 40 thick skeleton core flush door	22	no.			
H	Hole through 50 softwood for small pipe	15	no.			
I	63 upright mortice lock fixed to hardwood	35	no.			

The build-up of woodwork rates

Item A – Roof joists

Basic prices:

All-in rate for craftsman £4.50 per hour
Carcassing timber delivered to site £200 per cubic metre
Nails £1.50 per kg

Method

(a) *Materials*:

Cost of timber delivered to site per m^3	=	£200.00
Allow 2 hrs per m^3 for unloading at £4.00	=	8.00
		£208.00
Nails – 2 kg per m^3 at £1.50 per kg	=	3.00
		£211.00
Add 10% waste	=	21.10
		£232.10

Therefore cost of 150 × 50 per m = £232.10 × 0.15 × 0.05
= £1.74 . . . [1]

(b) *Labour*

Assume 1 joiner can fix 10 m of joist per hour

Therefore cost per m = $\frac{£4.50}{10}$ = 45p . . . [2]

Total rate ([1] + [2]) = <u>£2.19 per m</u>

Item B – Herringbone strutting

(*Note*: Herringbone strutting is measured in metres and is the horizontal distance taken over the joists. In practice, the work consists of two struts fixed diagonally between the joists which means that the quantity measured in the bill is **less than** the total amount of timber needed.)

Basic prices:

As above.

Method

(a) *Materials*:

Cost of timber including waste allowance = £232.10 per m^3

Therefore cost of 50 × 50 per m = £232.10 × 0.05 × 0.05
= 0.58

Length required per m as measured in BQ:

Assume 175 × 50 joists at 450 centres

– space between joists:

= 450 – (2 × 25)

= 400 mm

Diagonal distance from top of joist to bottom of adjacent joist:

$X^2 = 400^2 + 175^2$

$x = 0.16 + 0.031$

$= 437 \times 2 = 874$ mm

which means that for every 450 mm measured, 874 mm of timber are required

Therefore length required per m in BQ

$= \frac{1}{0.450} \times 0.874 = 1.94$ m

Therefore material cost per m = 1.94 × 58p
= £1.13 . . . [1]

Nails – a small nominal sum of, say, 10p per m would be sufficient . . . [2]

(b) *Labour*:

Assume 1 joiner can cut and fix 5 m of strutting per hour (as measured)

Therefore cost per m $= \frac{£4.50}{5} = 90p$ . . . [3]

Total rate ([1] + [2] + [3]) = £2.13 per m

Item C – Tongued and grooved flooring

(Note: The 150 mm dimension is 'nominal', i.e. measured **over** the tongue. After laying, the effective or 'laid' width of the board is the nominal width minus the tongue, i.e. 150 – 6 mm = 144 mm. This becomes important when calculating the total length of boards required per square metre of flooring.)

Basic prices:

Labour rates as above
150 × 25 T & G flooring £115.00 per 100 m
60 mm cut steel floor brads £1.00 per kg (approx. 250 per kg)

Method

(a) *Materials*:

Cost of flooring delivered to site per 100 m	=	£115.00
Allow ¼ hr per 100 m for unloading at £4.00	=	1.00
		£116.00
Nails – 5 per m = 500 per 100 m or 2 kg at £1.00	=	2.00
		£118.00
Add, say, 5% waste	=	5.90
		£123.90 per 100 m

Length required per $m^2 = \frac{1,000}{144} = 6.94$ m

Therefore cost per $m^2 = \frac{£123.90 \times 6.94}{100}$

$= £8.58$ . . . [1]

(b) *Labour*:

Assume 1 joiner can lay 15 m of boards per hour

Therefore cost per $m^2 = \frac{£4.50 \times 6.94}{15} = £2.08$ . . . [2]

Total rate ([1] + [2]) = £10.66 per m^2

Item D – Softwood skirting including grounds

Basic prices:

150 × 25 skirting £1.20 per metre
25 × 12 softwood grounds £15.00 per 100 m
Plugs (say) 4p each

Method

(a) *Materials*:

(1) 150 × 25 skirting

Cost of skirting delivered to site per metre	= £1.20
Unloading, stacking, etc., say	= 0.02
Nails (say)	= 0.10
	£1.32
Add 7½% waste	0.10
	£1.42

(2) 25 × 12 grounds

Cost of grounds delivered to site per 100 m	= £15.00
Unloading and stacking, say 3/20 hr per 100 m	= 0.60
	£15.60
Add 5% waste	= 0.78
	£16.38

Therefore cost per m = $\frac{£16.38}{100}$ = 16p

(3) Plugs

3⅓no. per m required at 4p each	= 0.13
Add 5% waste	= 0.01
	0.14

Total material cost = £1.72 . . . [1]

(b) *Labour*:

(1)	Skirting – 10 m per hour at £4.50	= 0.45
(2)	Grounds – 12 m per hour at £4.50	= 0.38
(3)	Plugging – 6 m per hour at £4.50	= 0.75
		£1.58 . . . [2]

Total rate ([1] + [2]) = <u>£3.30 per m</u>

Item E – Softwood window

Basic prices:

1,225 × 1,073 window complete £40.00 each

Method

(a) *Materials*:

1,225 × 1,073 window delivered to site	= £40.00 each
Add 2½% waste (to cover breakages, theft)	= 1.00
	£41.00 ... [1]

(b) *Labour*:

Craftsman unloading, hoisting and fixing ¾ hr with 50 per cent attendance by labourer

0.75 hr craftsman at £4.50	= £3.38
0.38 hr labourer at £4.00	= 1.52
	£4.90 ... [2]

Total rate ([1] + [2]) = £45.90 each

Item F – Plywood roof decking

Basic prices:

18 mm plywood in 2,240 × 1,220 sheets £6.20 per m²
Nails £1.50 per kg

Method

(a) *Materials*:

Cost of plywood delivered to site	= £6.20 per m²
Unload and stack (2 labs, 5 mins per sheet – approx. 3 m²) = 1/6 hr per 3 m²	= 0.22
Nails 4 rows per sheet at 200 mm c/s = 45 nails (approx. 0.2 kg) 0.2 kg at £1.50	= 0.30
	= £6.72
Add 5% waste	= 0.34
	£7.06 ... [1]

(b) *Labour*:

Assume 1 joiner can handle and fix 1 sheet in ⅓ hr with 25 per cent labourer's assistance

0.33 hr at £4.50	= £1.50
0.08 hr at £4.00	= 0.32
	= £1.82 per sheet

Therefore cost per m² = $\frac{£1.82}{3}$ = 61p ... [2]

Total rate ([1] + [2]) = £7.67 per m²

Item G – 40 mm thick internal flush door

Basic prices:

838 × 1,981 × 40 skeleton core flush door £25.00 each

Method

(a) Materials:

838 × 1,981 door delivered to site	= £25.00 each	
Unloading and storing say, 33p	= 0.33	
	£25.33	
Add 2½% waste (to cover damage and theft)	= 0.63	
	£25.96	. . . [1]

(b) *Labour*:

Assume 1 joiner can hang 40 mm flush door in 1 hr at £4.50 = £4.50 . . . [2]

Total rate ([1] + [2]) = £30.46 each

Item H – Hole for small pipe through 50 mm softwood

Labour only item – Assume 1 joiner can mark and drill hole through 50 mm softwood in 5 mins at £4.50 = 38p each.

Item I – 63 mm upright mortice lock

Basic prices:

Mortice lock £4.50 each (including screws)

Method

(a) *Materials*:

63 mm mortice lock at £4.50 each	= £4.50	
Add 2½% waste (to cover for losses, theft, etc.)	= 0.11	
	= £4.61	. . . [1]

(b) *Labour*:

Assume 1 joiner can cut out for and fix mortice lock in 1 hr

Add 50% for fixing to hardwood = 1½ hrs

1.5 hrs at £4.50 = £6.75 . . . [2]

Total rate ([1] + [2]) = £11.36 each

7.7 Plumbing and mechanical engineering installations

The work measured in this section of the bill of quantities can be divided into six broad categories, namely:

1 Gutterwork.
2 Pipework.
3 Ductwork.
4 Equipment and ancillaries.
5 Insulation.
6 Sundries and builders' work.

The measurable items in each of the above headings are further classified in accordance with the nature of the work as defined in Clause R.4.

Work carried out in this work section is executed by skilled operatives and concerns, in the main, the installation of pipework, ductwork and associated appliances and equipment necessary for the provision of services of one kind or another to the building.

The plumbing industry operates on a different basis from the majority of building trades, and employees are not therefore governed by the National Working Rule Agreement. The nature of the work is often highly skilled and this is reflected in payment of higher wages generally, together with enhanced terms and conditions of employment.

There are three recognised grades of operatives in the plumbing industry – the 'technical plumber', 'advanced plumber' and 'trained plumber', the highest paid of whom is the technical plumber whose basic rate is approximately half as much again as that of his skilled counterpart in other trades.

When pricing plumbing installations, therefore, it is important to establish the level of skill demanded by the work which in turn will determine the extent of the labour cost involved. Often a skilled plumber will work with a 'mate' who is normally an apprentice learning the trade and who provides assistance by carrying out general tasks such as helping to hoist and fix materials into position, fetching and carrying, making good, etc. Whether or not a mate will be employed will depend upon the nature and extent of the works, although those situations where he is deemed necessary must be identified at tender stage in order that his cost may be incorporated into the bill rates by way of establishing hourly 'gang' rates.

Factors affecting cost

Those factors which influence the cost of plumbing work must be appreciated by the estimator and may be identified as follows:

1 The rules for the measurement of the work.
2 The kind and quality of the materials specified.
3 The method of fixing.
4 The location of the work.
5 The method of jointing of pipework and fittings.

The nature of the work and the kind and quality of materials to be used will be clearly stated in the bill, and B.S. references and manufacturers' catalogue reference numbers given where appropriate to identify exactly the requirements of the architect, and in this respect the estimator would have no difficulty in establishing the material cost.

The rules for the measurement of plumbing work indicate that the quantities stated in the bill for gutterwork and pipework should be slightly higher than the actual corresponding lengths shown on the drawings. This is because the work is measured generally in metres taken **over** all associated fittings such as angles, branches, bends, etc., with such fittings then being enumerated and measured as **extra over** the gutter or pipe in which they occur (see Clauses R7.1, R10.1, etc.). In order to be strictly accurate when pricing for these fittings, therefore, the cost of providing and fixing them should first be calculated and then an amount equal to the cost of the short length of gutter (or pipe) which is displaced by the fitting **deducted**, thereby producing the **extra over cost** (whether this exercise is carried out in practice will of course depend upon the significance of the small saving in cost in relation to the tender figure as a whole).

The majority of the pipework, fittings and appliances required in any installation will require fixing rigidly in position whilst at the same time making due provision for any possible expansion and contraction which may take place. In recognition of this requirement, it can be seen that the method of fixing materials, whether by screws, nails, clips, bolts, etc., and the nature of the background to which items are fitted, i.e. timber, brick, concrete, etc., will determine a significant proportion of the total labour cost with the fixing of associated brackets, clips and other supports often taking longer than the assembling, jointing and fixing of the pipework itself.

The Standard Method recognises that the method of fixing and the type of background represent a major contribution to cost and therefore demands that such information be incorporated into the bill description.

Adjacent lengths of gutters, pipes and ductwork require fixing together in some way with the various methods used being dependent upon the materials specified and the level of performance demanded from the installation. Joints may be of the compression or capillary type for use with copper pipes, solvent-welded or similar for plastics, screwed joints for iron pipes, and so on. However. since the making of a secure joint is time-consuming, it follows that the labour cost of installing pipework depends not only on the length measured in the bill but also on the total number of joints in the whole layout. The number of joints, however, cannot be calculated in strict proportion to the total length of pipework since they will be dependent upon whether the layout comprises long uninterrupted lengths or a complicated system of short runs serving individual compartments or cubicles where the ratio of joints to pipework will be much greater.

Unfortunately, despite their importance with regard to cost, the rules of the Standard Method do not provide for the number of joints to be identified in the bill measurement or description except in the case of special or ornamental type connectors, all remaining joints and jointing materials being 'deemed to be included'. It is therefore the responsibility of the estimator to make the appropriate allowances in his calculations.

In an attempt to provide the estimator with some indication as to the complexity of the layout and the conditions under which the work is carried out (e.g. in open or confined spaces), the Standard Method requires that the location of the work should be stated together with a description of the installation where it is considered that drawings are inadequate.

Waste

As with all materials, a certain amount of wastage will be encountered, the extent of which is dependent upon:

1 The amount of cutting and discarding of short lengths which occur and which itself is dependent upon the layout of the installation as discussed above.
2 The amount of care taken in handling and fixing items, particularly vitreous china appliances which may easily be cracked or broken with rough handling resulting in considerable cost to the contractor.

7.7.1 Plumbing installations – guide to labour outputs

(All outputs are based on a plumber and mate, although the decision as to whether to employ a mate or not will depend upon the nature and extent of the work.)

(a) *Gutterwork (upvc)* 100 mm dia. half-round gutter on brackets screwed to timber Angle, outlet or stop end	 0.50 hrs per 4 m, length 0.20 hrs each
(b) *Rainwater pipework* (upvc) 68 mm dia RWP fixed with holderbats plugged to brickwork Bend, offset, single branch or shoe Add 10% for fire resistant sheet material Add 20% for cast iron	 0.30 hrs per 4 m length 0.15 hrs each

(c) (1) Service and waste pipework (upvc and polythene)

	Nature of background		
	Laid in trench	*Fixed to timber*	*Fixed to brickwork*
110 mm dia S. & V.P. V.P.	—	—	0.75 hr per 2 m length
Holderbat	—	—	0.20 hr each
Bend or single junction	—	—	0.40 hr each
Double junction	—	—	0.75 hr each
19 mm dia. waste/service pipe	0.02 hr per m	0.15 hr per m	0.17 hr per m
38 mm dia. waste/service pipe	0.03 hr per m	0.18 hr per m	0.20 hr per m
51 mm dia. waste/service pipe	0.04 hr per m	0.27 hr per m	0.30 hr per m
Tee	0.15 hr each		
Elbow	0.10 hr each		
Plastic trap	0.25 hr each		

Plastic pipes generally are jointed by a rubber ring push fit system or by solvent welded joints.

Soil and ventilation pipework with plain or socketed ends is sold in 2, 2.5, 3 and 4 m lengths with brackets and fittings purchased separately.

Polythene tubing is manufactured in low- and high-density grades with each grade composed of two classes of pipe. Pipework can be purchased by the metre, though for large contracts coils of 50, 100 and 150 m may be obtained.

(c) (2) Service and waste pipework (copper pipework and fittings)

Copper tube	*Fixed to timber*	*Fixed to brickwork*
15 mm dia.	0.27 hr per m	0.30 hr per m
22 mm dia.	0.30 hr per m	0.33 hr per m
35 mm dia.	0.45 hr per m	0.50 hr per m
Made bends –	15 mm dia. 0.20 hr each	
	22 mm dia. 0.27 hr each	
	35 mm dia. 0.54 hr each	

Fittings	*Compression*	*Capillary*
Straight coupling or bend	15 mm dia. 0.20 hr each	0.16 hr each
	22 mm dia. 0.24 hr each	0.19 hr each
	35 mm dia. 0.40 hr each	0.32 hr each
Tee	15 mm dia. 0.33 hr each	0.26 hr each
	22 mm dia. 0.40 hr each	0.32 hr each
	35 mm dia. 0.75 hr each	0.60 hr each
Brass stop cock	0.75 hr each	
Ball valve	1.00 hr each	
CP pillar tap	1.00 hr each	

Copper tubing is manufactured in three grades in accordance with B.S. 2871, Tables X, Y and Z, with diameters ranging from 15 to 67 mm. Table X tubing is available in 3 and 6 m lengths; Table Z in 6 m lengths only; and Table Y is 20 m coils. In addition, Table X and Y tubing is supplied coated with pvc for use in gas installations.

The method of jointing copper pipework is by capillary or compression fittings.

(d) Sanitary appliances

Suggested outputs include for unloading, storing, unpacking, getting into position and returning packing cases where necessary, in addition to the time taken for fixing.

SS single sink and drainer (to base unit)	0.75 hr each
Standard glass fibre bath with waste outlet	1.50 hr each
Basin pattern urinal with flushing cistern	2.75 hr each
Low-level wc suite complete	1.75 hr each
Lavatory basin with pedestal, incl. waste outlet	1.50 hr each

(e) Insulation

Sectional cylinder jacket in glass fibre fixed with steel straps for 1,000 mm high cylinder	1.25 hr each
25 mm thick expanded polystyrene rigid slab in insulating set to 4no. sides and top of cw cistern size 750 × 600 × 600 mm	0.30 hr
25 mm thick pre-formed glass fibre pipe insulation for 19 mm dia. pipe	0.15 hr per m

The typical bill items to be priced in this section are as follows:

					£	p
A	112 mm diameter half-round upvc gutter on and including brackets screwed to timber	115	m			
B	Extra for nozzle outlet	25	no.			
C	64 mm diameter upvc rainwater pipe fixed with and including brackets to brickwork	90	m			
D	Extra for 112½° offset bend	32	no.			
E	40 mm diameter upvc waste pipe with solvent-welded joints fixed with plastic clips to brickwork	200	m			
F	15 mm diameter copper tubing in cold water service with capillary joints fixed with copper clips to brickwork	174	m			
G	Extra for made bend	42	no.			
H	Extra for 15 × 15 × 15 tee	27	no.			
I	610 × 510 mm vitreous china wash hard basin with plug, chain and stay, Cat. Ref. No. – on and including pedestal Cat. Ref. No. – complete with pair 15 mm CP pillar taps and 32 mm CP waste fitting	15	no.			
J	50 mm thick glass fibre pipe insulation to 15 mm diameter copper pipe	100	m			

The build-up of plumbing rates

Item A – 112 mm dia. half-round pvc gutter

Assumptions

1 The gutter is supplied in 4 m lengths.
2 A single 4 m length of gutter requires support brackets at 1 m intervals and a union bracket at the junction of adjacent lengths.

Basic prices:

All-in rate for plumber £6.00 per hour
All-in rate for plumber's mate £4.00 per hour
112 mm dia. pvc gutter £7.00 per 4 m length
Support brackets 40p each
Union bracket 80p each

Method

(a) *Materials*:

Cost of 4 m length of gutter delivered to site	= £7.00
3no. support brackets at 40p each	= 1.20
1no. union bracket at 80p each	= 0.80
Screws (say)	= 0.15
	= £9.15
Allow 5% for unloading and waste	= 0.46
	£9.61 . . . [1]

(b) *Labour*:

Assume 1 plumber can set out and fix 4 brackets and clip 1no. 4 m length of gutter into position including any necessary cutting in $\frac{1}{2}$ hr

Therefore cost per length = £6.00 × $\frac{1}{2}$ hr = £3.00 . . . [2]

Therefore total cost per 4 m length ([1] + [2]) = £12.61

Total rate per m = $\frac{£12.61}{4}$ = £3.15

Item B – Extra for 112 dia. nozzle outlet

Assumptions

1 The nozzle outlet is 250 mm long.

Basic prices

112 dia. nozzle outlet £1.30 each

Method

(a) *Materials*:

Cost of 112 dia. nozzle outlet	= £1.30 each
1no. union bracket at 80p	= 0.80
Screws, say	= 0.05
	= £2.15
Allow 5% unloading and waste	= 0.11
	£2.26 . . . [1]

(b) *Labour*:

Assume 1 plumber can fix 1 bracket and a nozzle outlet in $\frac{1}{6}$ hr

$\frac{1}{6}$ hr at £6.00 = £1.00 . . . [2]

Total rate ([1] + [2]) = £3.26 each

Deduct

Cost of short length of gutter displaced by the nozzle outlet, i.e. £2.89 × 0.250 = 72p

Therefore extra over rate = £3.26 − 0.72
= £2.54 each

Item C – 64 dia. pvc rainwater pipe

Assumptions

1 The rainwater pipe is supplied plain ended in 4 m lengths with pipe connectors being used to form 'push fit' joints between adjacent lengths.
2 The rainwater pipe is supported at 2 m intervals and at every offset.

Basic prices:

64 dia. pvc rainwater pipe £7.20 per 4 m length
Connectors 70p each
Pipe brackets 70p each

Method

(a) *Materials*:

Cost of 4 m length of rainwater pipe	= £7.20	
1no. connector at 70p each	= 0.70	
2no. pipe brackets at 70p each	= 1.40	
Plugs, screws, etc., say	= 0.10	
	£9.40	
Allow 5% unloading and waste	= 0.47	
	£9.87	. . . [1]

(b) *Labour*:

Assume 1 plumber can set out and fix 2no. brackets (including plugging and screwing to brickwork) and 1no. 4 m length of rainwater pipe in 20 mins

$\frac{1}{3}$ hr at £6.00 = £2.00 . . . [2]

Total rate per 4 m length ([1] + [2]) = £11.87

Therefore total rate per m = $\frac{£11.87}{4}$ = £2.97

Item D – Extra for 112½° offset bend

Assumptions

1 The offset bend is 200 mm long.

Basic prices:

112½° offset bend £1.10 each

Method

(a) *Materials*:

Cost of $112\frac{1}{2}°$ offset bend	= £1.10 each
1no. connector at 70p	= 0.70
1no. pipe bracket at 70p	= 0.70
Plugs, screws, say	= 0.05
	£2.55
Allow 5% unloading and waste	= 0.13
	£2.68 . . . [1]

(b) *Labour*:

Assume 1 plumber can fix 1 bracket and a bend in 10 mins

$\frac{1}{6}$ hr at £6.00 = £1.00 . . . [2]

Total rate ([1] + [2]) = £3.68 each

Deduct:

Cost of short length of pipe displaced by the bend, i.e. £2.97 × 0.2 = 60p

Therefore extra over rate = £3.68 − 0.60 = £3.08 each

Item F – 40 mm dia. waste pipe

Assumptions:

1 The pipe is supplied in 4 m lengths and is plain ended, the method of jointing being by solvent-welded sockets.

2 The pipe is supported by pipe brackets at 750 mm centres horizontally and 1,250 mm centres vertically. However, since no distinction is made in the SMM an **average** fixing interval of 1 m is assumed.

3 Due to the cutting required, assume 2 sockets will be needed, on average, to every 3 m length of pipe.

Basic prices:

40 mm dia. pvc pipe £3.00 per 3 m length
Pipe clips 20p each
Double socket 40p each

Method

(a) *Materials*:

Cost of 40 mm pvc pipe per 3 m length	= £3.00
3no. pipe brackets at 20p each	= 0.60
2no. double sockets at 40p each	= 0.80
Solvent cement, say	= 0.03
	£4.43
Allow 5% unloading and waste	0.22
	£4.65 . . . [1]

(b) *Labour*:

Assume gang cost of plumber and mate at £6.00 + £4.00 = £10.00 per hour

Assume plumber and mate can fix a 3 m length (including 2 joints), brackets plugged and screwed to brickwork in ½ hr

½ hr × £10.00 = £5.00 . . . [2]

Total rate per 3 m length ([1] + [2]) = £9.65

Therefore rate per m = $\frac{£9.65}{3}$ = £3.22

Item G – 15 dia. copper pipe in cold water service

Assumptions

1 The copper tubing is supplied in 3 m lengths.
2 The pipe is supported at 1,000 mm intervals by copper fixing clips.
3 Assume 1 straight coupling per 3 m length = ⅓rd coupling per m of pipe.

Basic prices:

15 dia. copper tube in 3 m lengths £56.00 per 100 m
Straight coupling 75p each
Pipe clips 15p each

Method

(a) *Materials*:

Cost of 15 dia. copper tube per 3 m length =	$\frac{£56 \times 3}{100}$	= £1.68
1no. straight coupling at 75p each		= 0.75
3no. pipe clips at 15p each		= 0.45
Plugs and screws, say		= 0.15
Jointing materials (flux, steel wool, etc.), say		= 0.05
		£3.08
Allow 5% unloading and waste		= 0.16
		£3.24 . . . [1]

(b) *Labour*:

Assume gang cost of £10.00 per hour as before
Assume gang can set out and fix 3 pipe clips plugged and screwed to brickwork, 3 m length of tubing and 1 coupling in ¾ hr × £10.00 = £7.50 . . . [2]

Total rate per 3 m length ([1] + [2]) = £10.74

Therefore rate per m = $\frac{£10.74}{3}$ = £3.58

Item H – Made bend on 15 dia. copper tube

No materials required – labour only item.

Labour:

Assume gang cost of £10.00 per hour as before
Assume gang can form made bend in 10 mins
Total rate = $\frac{1}{6}$ hr × £10.00 = <u>£1.67 each</u>

Item I – 15 × 15 × 15 dia. copper tee

Assumptions

1 Although fittings are measured 'extra over' the length of pipe in which they occur, the amount of pipe displaced by the tee in this case is negligible and therefore ignored in the calculation.

Basic prices:

15 dia. copper tee £1.40 each

Method

(a) *Materials*:

Cost of 15 dia. copper tee £1.40 each	= £1.40	
Allow 2½% unloading and waste	= 0.04	
Jointing materials (as before), say	= 0.02	
	£1.46	. . . [1]

(b) *Labour*:

Assume gang cost of £10.00 per hour as before
Fixing a tee involves making 3no. joints:
Assume gang can fix 1 tee in ¼ hr
¼ hr at £10.00 = £2.50 . . . [2]

Total rate ([1] + [2]) = <u>£3.96 each</u>

Item J – 610 × 510 mm vitreous china wash hand basin and pedestal with pair of 15 mm CP pillar taps and 32 mm CP waste fitting

Basic prices:

610 × 510 mm wash hand basin and pedestal £45.00 each
15 dia. CP taps £6.00 each
32 dia. CP waste fitting £3.50 each

Method

(a) *Materials*:

Cost of 610 × 510 mm wash hand basin and pedestal	= £45.00	
2no. 15 dia. CP taps at £6.00 each	= 12.00	
32 mm CP waste fitting at £3.50 each	= 3.50	
Jointing materials (bedding waste, basin on pedestal, etc.), say	= 0.20	
Screws, say	= 0.10	
	£60.80	
Allow 2½% unloading and waste	= 1.52	
	£62.32	. . . [1]

(b) *Labour*:

Assume gang cost of £10.00 per hour as before
Assume gang can assemble and fix basin, pedestal, taps, etc. in $1\frac{1}{2}$ hrs
$1\frac{1}{2}$ hrs × £10.00 = £15.00 . . . [2]

Total rate ([1] + [2]) £77.32 each

Item K – 50 mm thick glass fibre pipe insulation to 15 dia. copper pipe

Basic prices:

4 m long × 75 mm wide glass fibre pipewrap
50 mm thick £1.25 per roll

Method

(a) *Materials*:

Cost of 4 m length of pipewrap	= £1.25	
Allow 5% laps, waste and ties	= 0.06	
	£1.31	. . . [1]

(b) *Labour*:

Assume gang cost of £10.00 per hour as before
Assume gang can fix 1no. 4 m length in $\frac{1}{4}$ hr (work probably executed in confined areas, roof spaces, etc.)
$\frac{1}{4}$ hr at £10.00 = £2.50 . . . [2]

Total rate per 4 m roll ([1]) + [2]) £3.81

Therefore rate per m = $\frac{£3.81}{4}$ = 95p

7.8 Floor, wall and ceiling finishes

The floor, wall and ceiling finishes section comprises a wide variety of skills and materials, the work often being carried out by specialist sub-contractors covering:

1 *In-situ* finishings.
2 Tile, slab and block finishings.
3 Flexible sheet finishings.
4 Dry linings and partitions.
5 Suspended ceilings.
6 Fibrous plaster.
7 Fitted carpeting.

7.8.1 *In-situ* finishings

Covers mainly (a) plasterwork to walls and ceilings and (b) backings and floor screeding.

Plasterwork generally is measured and priced in square metres; however, the plaster itself of whatever type, together with sand, lime and cement, are all purchased by weight with the plaster, lime and cement supplied in 50 kg bags, the material cost therefore being calculated in terms of kg per square metre.

Modern plasters are manufactured from gypsum rock crushed to a fine powder with various additives included to modify setting times and workability and to produce plasters with lightweight, insulating or hard-wearing characteristics. Most plasters can be used neat and require only to be mixed with clean water on site to render the material ready for use.

A number of different types of plaster are available to suit all kinds of backgrounds at various levels of performance as indicated in Table 7.1.

Table 7.1 Plasters for internal use

Type	Thickness (mm)	Mix	Coverage in m² per tonne	Properties and uses
Carlite undercoat	11	Neat plaster	130 to 150	Brick and block walls
Carlite finish	2	Neat plaster	410 to 500	Brick and block walls
Thistle undercoat	11	1 : 3 plaster and sand on brick walls	234 to 240	Floating coat scratched for key
		1 : 2 plaster and sand on concrete and lightweight blockwork	175 to 185	
Thistle finish	2	Neat	390	
Thistle board finish	5	Neat	165	Finishing coat on plasterboard base
Sirapite (finish only)	3	Neat for a smooth finish or as a 1 : 1 mix with sand for a textured finish	250 to 270	Used on most plaster and cement and sand undercoats

Coverages quoted include an allowance for shrinkage and consolidation during mixing and application. A further allowance of, say, 5–10 per cent should be made for wastage of materials.

Labour

The work involved in *in-situ* plastering is normally undertaken by 'gangs' comprising 2 plasterers and 1 labourer or 3 plasterers and 2 labourers, the unskilled operatives being responsible for the mixing (by hand) and transporting of materials to the plasterers.

Floor screeds

Floor screeds and backings for wall tiles, etc. comprise a mixture of cement and sand in various proportions to which may be added granite chippings to produce a particularly hard-wearing floor finish known as 'granolithic'.

The principles involved in pricing are similar to those adopted for mortar mixing, with the materials usually being mixed by machine with allowances to be made for loss of volume due to compression when laying in addition to the shrinkage and waste factors.

Labour

The floor screeding gang would normally comprise skilled and unskilled operatives in the ratio of 1 : 1

Guide to average labour outputs for *in-situ* work

Operation	*Output in skilled hours per m²*
6 mm skim coat gypsum plaster on:	
walls	0.35
soffits	0.58
13 mm two-coat lightweight plaster on:	
walls	0.90
soffits	1.15
13 mm three-coat lightweight plaster on:	
metal lath walls	1.10
metal lath soffits	1.32
Angle bead, incl. working in two coats lightweight plaster	0.15 per m
25 mm cement and sand screed (1 : 3)	0.30
50 mm cement and sand screed (1 : 3)	0.45
Steel trowelled finish	0.20
Finishing to falls or crossfalls	0.25
25 mm granolithic bed (1 : 2½)	0.45
50 mm granolithic bed (1 : 2½)	0.65

(Note: Where access is difficult or where work is executed in narrow widths (n.e. 300 mm wide), the following additions to the labour cost should be included:
(a) Work to staircase areas, ceilings and beams over 3.5 m high and in compartments n.e. 4 m² on plan +25%
(b) All work n.e. 300 mm wide +30%)

7.8.2 Tile, slab and block finishings

The unit of measurement is again the square metre, with work executed in narrow widths n.e. 300 mm being given separately and skirtings and sills measured and priced in metres.

Tiles, slabs and blocks for walls and floors are manufactured in a wide range of materials, textures, colours and prices, and in view of this the Standard Method demands that a considerable amount of information is given in the bill description in order for the estimator to be able to identify exactly what is required.

Tiles are purchased by the square metre, in units of 10, 100 or whatever quantity fits into easily-managed packs depending on individual suppliers. The size of the tile and the thickness of the joint must be known so that the number required per square metre can be determined.

In addition to pricing for the tiles, the unit rate must also include for the cost of the bedding and jointing material which may include cement and sand, coloured mortar or adhesives of one sort or another as recommended by the supplier and specified in the bill description.

Guide to average outputs for tile, slab and blockwork

Operation	*Output in skilled hours per m²*
108 × 108 × 4 mm glazed wall tiles fixed with adhesive and pointed in white cement	1.30
152 × 152 × 6 mm glazed wall tiles fixed with adhesive and pointed in white cement	1.05
152 × 152 × 16 mm quarry tile flooring laid level and to falls (cement and sand bed not incl.)	0.75
As above, laid to falls crossfalls and slopes	0.80
152 × 76 × 16 mm cove skirting	0.30 (per m)
305 × 305 × 25 mm precast terrazzo tile paving pointed in white cement	1.00
300 × 300 × 2.5 mm pvc floor tiles fixed with adhesive	0.40

(Note: The same allowances as for *in-situ* work should be made where work is executed in narrow widths or in confined or difficult locations. An allowance of approximately 5 per cent should be made for cutting and waste.)

7.8.3 Sheet finishings

Examples of work measured under this heading would include sheet finishings such as plastic and plasterboard fixed to walls, ceilings, columns, etc. for subsequent finish with a skim coat of plaster or direct decoration with paint.

The kind, quality and thickness of the material is to be stated, together with the method of fixing and treatment of the joints.

Plasterboard, being a popular material for wall and ceiling backings, is normally fixed to a timber background with galvanised wire clout nails with joints between adjacent boards being filled with neat plaster and scrimmed with a strip of jute cloth or paper.

The unit of measurement is the square metre with work not exceeding 300 mm wide being so described.

The same allowances as before should be made where work is executed in confined areas not exceeding 4 m² on plan, in staircase areas or in narrow widths.

Allow approximately 5 per cent for cutting and waste.

The remaining categories of work in this section are normally carried out by specialist firms and are not considered here.

The typical bill items to be priced in this section are as follows:

					£	p
A	13 mm two-coat carlite plaster with steel trowelled finish to brick or block walls over 300 wide	731	m^2			
B	Not exceeding 300 wide	195	m^2			
C	6 mm skim coat of carlite plaster to plasterboard walls over 300 wide	640	m^2			
D	13 mm two-coat thistle plaster and sand (1 : 3) with steel trowelled finish to brick or block walls over 300 wide	272	m^2			
E	Galvanised metal angle beadfixed with plaster dabs including working 13 mm two coat finishing to both sides	850	m			
F	50 mm thick granolithic bed (1 : 2½) trowelled with a plain surface to level concrete floors over 300 wide in compartments not exceeding 4 m^2 on plan	1,109	m^2			
G	152 × 152 × 12 mm clay quarry tiles laid to a regular pattern, bedded in cement mortar (1 : 3) 12 mm thick and pointed with 6 mm joints straight both ways to level concrete floors over 300 wide	338	m^2			
I	152 × 152 × 6 mm white glazed ceramic tiles fixed to regular pattern with adhesive to plastered walls grouted with white cement over 300 wide	60	m^2			
H	150 × 75 × 12 mm clay skirting tiles with rounded top edge and coved bottom laid with 6 mm joints and bedded and pointed in cement mortar (1 : 3)	423	m			
J	12.7 mm thick plasterboard with square joints filled with plaster and scrimmed for ceiling backings fixed with galvanised clout nails over 300 wide	950	m^2			

The build-up of floor, wall and ceiling finishings rates

Item A – 13 mm two-coat carlite plaster to walls

Basic prices:

All-in rate for craftsman £4.50 per hour
All-in rate for labourer £4.00 per hour
Carlite undercoat (browning) plaster £120.00/tonne
Carlite finishing plaster £96.00 per tonne

Method

(a) *Materials*:

Undercoat:

Cost of undercoat plaster delivered to site per tonne	= £120.00
Allow 1 hr per tonne unload and stack	= 4.00
	£124.00

Coverage of undercoat plaster 11 mm thick is approx. 140 m² per tonne

Therefore cost per m² $= \dfrac{£124.00}{140} = 89p$

Finishing coat:

Cost of finishing plaster delivered to site per tonne	= £96.00
Allow 1 hr per tonne unload and stack	= 4.00
	£100.00

Coverage of finishing plaster 2 mm thick is approx. 450 m² per tonne

Therefore cost per m² $= \dfrac{£100.00}{450} = 22p$

Total material cost per m² =	£1.11	
Allow 5% waste	0.06	
	£1.17	. . . [1]

(b) *Labour:*

Based on 3 : 1 gang of 3 plasterers and 1 labourer. Hourly gang cost = (3 × £4.50) + £4.00 = £17.50

Assume output per skilled operative for two-coat work to walls is 0.9 hr per m²

Therefore cost per m² $= \dfrac{£17.50 \times 0.9}{3} = £5.25$ . . . [2]

Total rate ([1] + [2]) = <u>£6.42 per m²</u>

Item B – As Item A but in narrow widths, i.e. 300 mm

Method

(a) *Materials*:
The material cost per m^2 remains unchanged at £1.17 . . . [1]

(b) *Labour*:
Allow an additional 30 per cent for working in narrow widths
= £5.25 × 1.30 = £6.83 . . . [2]

Total rate ([1] + [2]) = £8.00 per m^2

Item C – 6 mm skim coat of plaster to plasterboard walls

Basic prices:

Thistle board finish plaster £102 per tonne

Method

(a) *Materials*:

Cost of board finish plaster delivered to site per tonne	=	£102.00
Allow 1 hr per tonne unload and stack	=	4.00
		£106.00

Coverage of board finish plaster 6 mm thick is approx. 165 m^2 per tonne

Therefore cost per m^2 = $\frac{£106.00}{165}$ = 64p

Allow 5 per cent waste 03p

Material cost per m^2 = 67p . . . [1]

(b) *Labour*:
Based on 3 : 1 gang. Hourly cost £17.50 as before assume output per skilled operative for 6 mm skim coat of plaster is 0.35 hr per m^2

Therefore cost per m^2 = $\frac{£17.50 \times 0.35}{3}$ = £2.04 . . . [2]

Total rate ([1] + [2]) = £2.71 per m^2

Item D – 13 mm two-coat thistle plaster and sand (1 : 3) to walls

Basic prices:

Thistle undercoat plaster £135.00 per tonne
Thistle finishing plaster £105.00 per tonne
Coarse sand £11.00 per tonne

Method

(a) *Materials*:

Undercoat:

Cost of thistle undercoat plaster delivered to site per tonne	= £135.00
Allow 1 hr per tonne unload and stack	= 4.00
	£139.00

Cost of (1 : 3) mix by volume:

1 m^3 thistle undercoat plaster costs 1.44 × £139	= £200.16
3 m^3 coarse sand cost 1.52 × £11.00 × 3	= 50.16
	£250.32

Coverage of undercoat 11 mm thick per tonne of plaster is approx. 240 m^2

Therefore cost per tonne of plaster = $\frac{£250.32}{1.44}$ = £173.83

Therefore cost per m^2 = $\frac{£173.83}{240}$ = 72p

Finishing coat:

Cost of thistle finishing plaster delivered to site per tonne	= £105.00
Allow 1 h per tonne unload and stack	= 4.00
	£109.00

Coverage of neat plaster finishing coat 2 mm thick is approx. 400 m^2 per tonne

Therefore cost per m^2 = $\frac{£109.00}{400}$ = 27p

Total material cost per m^2	= 99p	
Allow 5 per cent waste	0.05	
	£1.04	. . . [1]

(b) *Labour*:

Assume labour cost as before for 13 mm thick two-coat work at £5.25 per m^2 . . . [2]

Total rate ([1] + [2]) = £6.29 per m^2

Item E – Galvanised metal angle bead

Basic prices:

Galvanised angle bead 80p per 2.4 m length

Method

(a) Materials:

Cost of galvanised metal angle bead delivered to site	= 80p
Add 10 per cent unloading and waste	= 08
Allow for neat plaster for fixing, say	= 10
	98p

Therefore cost per m $= \frac{98}{2.4} = 41\text{p}$ . . . [1]

(b) *Labour*:

Based on gang cost of £17.50 per hour as before assume skilled operative can fix 7 m of angle bead per hour

Therefore cost per m $= \frac{£17.50}{(7\text{ m} \times 3)} = 83\text{p}$ . . . [2]

Total rate ([1] + [2]) = £1.24 per m

Item F – 50 mm thick granolithic flooring is compartments n.e. 4 m² on plan

Basic prices:

6 mm dust granolithic chippings £30 per tonne
Cement £64.00 per tonne
All-in rate for 5/4½ mixer, including fuel, oil, etc., £1.20 per hour

Method:

(a) *Materials*:

Cost of cement delivered to site per tonne	= £64.00
Allow 1 hr per tonne unload and stack	= 4.00
	£68.00

Cost of (1 : 2½) mix by volume:

1 m³ cement costs £68.00 × 1.44	=	£ 97.92
2½ m³ grano chippings cost 1.35 × £30.00 × 2½	=	101.25
	=	£199.17
Add 30 per cent shrinkage and consolidation	=	59.75
		£258.92
Add 5 per cent waste	=	12.95
		£271.87

Therefore cost per m³ $= \frac{£\,271.87}{3\frac{1}{2}} = £77.68$

Cost of Mixing:

All-in rate for 5/3½ mixer per hour	= £1.20
Operator at £4.00 + 10p (NWRA)	= 4.10
	£5.30

Assume output of mixer is 1 m³ per hour
Therefore cost of mixing per m³ = £5.30

Total cost of mixed materials = £82.98 per m³

Therefore cost per m², 50 mm thick	= £82.98 × 0.05	
	= £4.15	
Allow for screed laths, say	= 0.20	
	£4.35	. . . [1]

(b) *Labour*:

Based on gang cost of £17.50 per hour as before

Assume skilled operative can lay 1 m² in	0.65 hr
Add for steel trowelled finish	0.20 hr
	0.85 hr

Therefore labour cost per m² = $\frac{£17.50 \times 0.85}{3}$ = £4.96

Add 25 per cent for work in confined areas = £1.24

£6.20 . . . [2]

Total rate ([1] + [2]) = £10.55 per m²

Item G – 152 × 152 × 6 mm white glazed wall tiling

Basic prices:

152 × 152 × 6 mm wall tiles £18.00 per 100
Ceramic tile adhesive £3.00 per litre

Method

(a) *Materials*:

Cost of glazed wall tiles delivered to site per 100	= £18.00
Allow 5 per cent unloading and waste	= 0.90
	£18.90

No. of tiles required per m² of wall = $\frac{1,000}{152} \times \frac{1,000}{152}$ = 43

Therefore cost per m² = $\frac{£18.90 \times 43}{100}$ = £8.13

Allow ⅓ litre adhesive at £3.00 per l = 1.00

£9.13

Allow for white cement grouting, say 0.10

£9.23 . . . [1]

(b) *Labour*:

Assume skilled operative can fix 1 m^2 of 152 × 152 tiles, including grouting, in 1.05 hrs

Therefore cost per m^2 = £4.50 × 1.05

= £4.73 . . . [2]

Total rate ([1] + [2]) = £13.96 per m^2

Item H – 152 × 152 × 16 mm clay quarry tiled paving on 12 mm cement mortar bed

Basic prices:

152 × 152 × 16 mm clay quarry tiles £3.50 per 10

Sand £11.00 per tonne

Cement £64.00 per tonne

All-in rate for 5/3½ mixer, incl. fuel. £1.20 hour

Method:

(a) *Materials*:

Cement mortar bed:

Cost of cement delivered to site per tonne	=	£64.00
Allow 1 hr per tonne unload and stack	=	4.00
		£68.00

Cost of (1 : 3) mix by volume:

1 m^3 cement costs £68.00 × 1.44	=	£97.92
3 m^3 sand cost £11.00 × 1.52 × 3	=	50.16
		£148.08
Allow 30 per cent shrinkage and consolidation	=	44.42
		£192.50
Add 5 per cent waste	=	9.63
		£202.13

Therefore cost per m^3 $= \frac{£202.13}{4}$ = £50.53

Cost of mixing as before at £5.30 per m^3

Therefore cost of mixed mortar = £55.83 per m^3

Therefore cost per m^2, 12 mm thick = £55.83 × 0.012

= 67p . . . [1]

Cost of laying bed:

Based on 3 : 1 gang cost of £17.50 per hour as before

Assume 1 skilled operative can lay 1 m^2 in 0.2 hr

Therefore cost per m^2 $= \frac{£17.50 \times 0.2}{3}$

= £1.17 . . . [2]

Tiles:

Cost of 152 × 152 × 16 mm tiles delivered to site = £3.50 per 100

Allow 5 per cent unloading and waste = 0.18

£3.68

No. of 152 × 152 tiles required per m² of floor allowing for a 6 mm joint is

$$\frac{1{,}000}{(152 + 6)} \times \frac{1{,}000}{(152 + 6)} = 40$$

Therefore cost of tiles per m² = $\frac{£3.68 \times 40}{10}$ = £14.72

Allow for mortar for pointing, say 0.20

£14.92 . . . [3]

(b) *Labour*:

Assume skilled operative can lay and point up 1 m² of floor tiles in 0.75 hr

Therefore cost per m² = £4.50 × 0.75 = £3.38 . . . [4]

Total rate ([1] + [2] + [3] + [4]) = £20.14 per m²

Item 1–150 × 75 mm clay coved skirting

Basic prices:

150 × 75 × 16 mm clay coved skirting tiles £4.50 per 10

Method:

(a) *Materials*;

Cement mortar bed:

Quantity of mortar required per m of skirting

= 1.00 × 0.075 × 0.012 = 0.0009 m³

Cost of cement mortar per m³ from previous example £53.87

Therefore cost per m = £53.87 × 0.0009 = 5p . . . [1]

Tiles:

Cost of skirting tiles delivered to site per 10 = £4.50

Allow 5 per cent unloading and waste = 0.23

£4.73

No. of tiles required per m with 6 mm joints

$$= \frac{1{,}000}{(150 + 6)} = 6.33$$

Therefore cost of skirting tiles per m = $\frac{£4.73 \times 6.33}{10}$ = £2.99

Add for mortar for pointing, say 0.02

£3.01 . . . [2]

(b) *Labour*:

Assume skilled operative can fix 1 m of coved skirting in 0.30 hr, including bedding and pointing in cement mortar

Therefore cost per m = £4.50 × 0.30 = £1.35 . . . [3]

Total rate ([1] + [2] + [3]) = <u>£4.41 per m</u>

Item J – 12.7 mm plasterboard to soffit

Basic prices:

12.7 mm plasterboard in 1,829 × 1,200 mm sheets £6.00 per sheet

Jute scrim cloth £4.00 per 100 m roll

20 mm galvanised wire clout nails £40 per 25 kg

Method:

(a) *Materials*:

Cost of 12.7 mm plasterboard delivered to site per sheet	= £6.00
Unload and stack $\frac{1}{10}$ hr at £4.00	= 0.40
	£6.40

Amount of scrim cloth required per sheet (taking half the perimeter)

= 1,829 + 1,200 mm = 3.03 m

Therefore cost of scrim cloth per sheet = $\frac{£4.00 \times 3.03}{100}$ = 12p

No. of nails required per sheet:

Assume joists are at 450 mm centres

Therefore no. of rows of nails = $\frac{1,829}{450}$ = 4 + 1 (end) = 5 rows

Assume nails are at 150 mm centres

Therefore no. of nails per sheet $\frac{1,200}{150}$ × 5 rows = 45no.

There are approx. 350 galvanised nails per kg

Therefore cost of nails per sheet

= $\frac{£40.00}{25} \times \frac{45}{300}$ = 21p

Total material cost per sheet	= £6.73	
Allow for neat plaster fixing scrim, say	0.10	
	£6.83	
Add 5 per cent waste	0.34	
	£7.17	. . . [1]

(b) *Labour*:

Based on 1 : 1 gang. Hourly gang cost = £4.50 + £4.00
= £8.50

Assume gang can fix 1,829 × 1,200 mm size sheet in 0.33 hr
Therefore cost per sheet = £8.50 × 0.33 = £2.81 . . . [2]

Total cost per sheet ([1] + [2]) = £9.98

$$\text{Total rate} = \frac{\pounds 9.98}{(1{,}829 \times 1{,}200)} = \frac{\pounds 9.98}{2.2\ \text{m}^2} = \underline{\pounds 4.54 \text{ per m}^2}$$

7.9 Glazing

The glazing section of the Standard Method is one of the shortest and most straightforward of all the categories with regard to the measurement and pricing of the work involved.

The majority of the glazing section, Clauses U.1 – U.13, deals with glass in openings such as windows and doors with the remaining clauses, U.14–U.18, covering the measurement of leaded and copper lights, mirrors, patent glazing and domelights.

Glass and glazing generally is measured in square metres, the exception being 'special glass' which is enumerated in accordance with Clause U.5. Where the former rule applies, the kind and quality of the glass is stated together with the actual area as fixed in position, although in some instances this information alone does not give a true indication of cost. For example:

1 The number of panes of glass making up the total area will have a significant effect upon the time taken to carry out the glazing.
2 The larger the number of panes making up the area, the longer the total perimeter of glass requiring putty and/or beads.

Thus the smaller the panes, the greater the cost per square metre, and in recognition of this the sizes of panes grouped in four different ranges must be given in the bill description in accordance with Clause U.4.1.

7.9 Materials

There are many different types of glass available in a number of thicknesses and possessing a wide range of properties such as toughened, wired, solar reflecting, etc. to suit varying uses. The main types of glass are specified in accordance with their method of manufacture, e.g. sheet, cast, float, wired, etc. and, in addition to plain glass, various patterns and tints are available. Since the price of glass can vary considerably, the kind, quality and thickness to be used must be clearly stated together with the method of glazing and securing the glass in position (i.e. putty and sprigs, pinned or screwed beads, clips, etc.) and the nature of the surround (usually wood or metal).

It is important to note that the cost of putty and sprigs is to be included in the rate per square metre for the glass whereas wood beads are measured and priced for separately in the woodwork section.

7.9.2 Waste

Normally, panes would be cut to size by the supplier from a schedule of measurements taken on site which reduces the risk of breakages when compared with cutting on site.

An allowance of 5 per cent waste is sufficient on pre-cut panes whereas 10 per cent should be allowed where cutting is performed on site.

7.9.3 Labour

Glazing is executed by skilled labour and may be performed by specialist glaziers although, perhaps on a smaller scale, plumbers and joiners often carry out this type of work. For extra large and heavy panes of glass, it should be remembered that two or more operatives may be required to manhandle and fix the glass in position. The location of the glazing work may affect the cost where access is difficult or where the work is carried out in confined spaces, and in this respect Clause U1 requires such information to be given where the tender documentation is insufficient.

Labours on glass, such as raking cutting, curved cutting, embossing, etc., are measured in metres or square metres depending on the nature of the work in accordance with Clause U.6.

Glazing – guide to labour outputs

Outputs for glazing 4 mm glass to wood with putty and sprigs – skilled hours per m²	
Size of panes	*Hours*
n.e. 0.10 m²	0.60
0.10–0.50 m²	0.55
0.50–1.00 m²	0.50
over 1.00 m²	0.45

Notes: (a) For 6 mm glass, increase outputs by 7½ per cent.
(b) Allow, in addition, 50 per cent of the above outputs where glass is cut on site.
(c) Allow an extra 10 per cent for glazing with pinned beads.
(d) Allow an extra 20 per cent for glazing with screwed beads.

Quantity of putty required per m² of glass		
Size of panes	*Glazed with putty and sprigs (Kg)*	*Glazed with beads (Kg)*
n.e. 0.10 m²	3	1
0.10–0.50 m²	2	¾
0.50–1.00 m²	1	⅓
over 1.00 m²	¾	¼

The typical bill items to be priced in this section are as follows:

				£	p
A	3 mm thick SQ quality sheet glass to wood frame with putty in panes not exceeding 0.10 m²	150	m²		
B	As A, in panes 0.10–0.50 m²	95	m²		
C	6 mm thick clear float glass GG quality to wood frame with screwed beads in panes 0.50–1.00 m²	45	m²		

The build-up of unit rates for glazing work

Item A – 3 mm glass to wood frame with putty in panes n.e. 0.10 m²
Basic prices:

3 mm SQ quality sheet glass £11.00 per m²
Putty 40p per kg

Method:

(a) *Materials*:

Cost of 3 mm sheet glass delivered to site per m²	=	£11.00
3 kg of putty at 40p per kg	=	£1.20
Sprigs for fixing, say	=	0.05
		£12.25
Add 5% unloading and waste	=	0.61
		£12.86 . . [1]

(b) *Labour*:

Assume glazier can glaze 1 m² of glass in panes n.e. 0.10 m² in 0.60 hr.
Therefore cost per m² = £4.50 × 0.60 = £2.70 . . . [2]

Total rate ([1] + [2]) = £15.56 per m²

Item B – As Item B, in panes 0.10–0.50 m²
Method:

(a) *Materials*:

Cost of 3 mm SQ quality sheet glass as before	=	£11.00
2 kg. of putty at 40p per kg	=	0.80
Sprigs for fixing, say	=	0.04
		£11.84
Add 5% unloading and waste	=	0.59
		£12.43 . . . [1]

(b) *Labour*:

Assume glazier can glaze 1 m^2 of glass in panes 0.10–0.50 m^2 in 0.55 hr

Therefore cost per m^2 = £4.50 × 0.55 = £2.48 . . . [2]

Total rate ([1] + [2]) = £14.91 per m^2

Item C – 6 mm float glass to wood frame with screwed beads in panes 0.50–1.00 m^2

Basic prices:

6 mm GG quality clear float glass £18.00 per m^2

Method:

(a) *Materials*:

Cost of 6 mm float glass delivered to site per m^2	=£18.00	
$\frac{1}{3}$ kg putty at 40p per kg	= 0.13	
	£18.13	
Add 5% unloading and waste	= 0.91	
	£19.04	. . . [1]

(beads and screws measured and priced in the 'Woodwork' work section)

(b) *Labour*:

Assume glazier can glaze 1 m^2 of 6 mm thick float glass in panes 0.50–1.00 m^2 in 0.50 hr + $7\frac{1}{2}$% = 0.54 hrs

Therefore cost per m^2 = £4.50 × 0.54	= £2.43	
Add 20% for glazing with screwed beads	= 0.49	
	£2.92	. . . [2]

Total rate ([1] + [2]) = £21.96 per m^2

7.10 Painting and decorating

The work contained in this section covers two broad categories, namely:

1 Painting, polishing and similar work.
2 Wallpapering and other similar wall coverings.

Painting and decorating work is largely labour-intensive with, on average, approximately 70 per cent of the total cost being attributable to this resource. It is essential, therefore, that the estimator is provided with the fullest possible information in connection with the conditions under which the work is to be carried out with the following work being specifically identified:

New work internally
New work externally
Repainting and redecoration work internally
Repainting and redecoration work externally
Redecoration and work of a restorative nature
Work carried out on members before fixing
Work to ceilings and beams over 3.5 m high

7.10.1 Measurement

Painting and decorating may be measured in square metres, metres or enumerated depending on the nature of the work with the area covered being stated, although voids not exceeding 0.5 m^2 in area are ignored and appropriate allowances made for mouldings, sinkings, corrugations, etc.

Generally the information given in the bill description includes the following:

1 The type of surface to be treated, e.g. walls, ceilings, skirtings, windows, etc.
2 The type of decoration work and number of coats to be applied.
3 The preparatory work required, e.g. rubbing down, burning off, etc.
4 The nature of the surface to be treated, e.g. plaster, brick, wood, metal, etc.

7.10.2 Materials

The quantity of materials required is shown in Table 7.2.

Table 7.2 Quantity of materials required per 100 m² (per coat)

Material	Nature of background					
	Plaster	Concrete	FF Brickwork	Blockwork	Wood	Metal
Emulsion paint (litres)	7½	8	10	12½	—	—
Thinned emulsion paint (mist coat) (litres)	15	16	20	25	—	—
Cement based paint (litres)	—	15½	22	22	—	—
Wood primer (litres)	—		—	—	10½	—
Metal primer (litres)	—	—	—	—	—	10
Undercoat (litres)	8	12½	14½	14½	9	9
Gloss finish (litres)	8	11	12½	14½	9	9
Oil-based eggshell (litres)	8	8	11	12½	8½	—
Knotting (litres)	—	—	—	—	¾	—
Glasspaper (sheets)	—	—	—	—	8	—
Putty (litres)	—	—	—	—	2½	—
Plastic coating (kg)	40	40	—	—	—	—
	(depending on type of textured finish)					
Polyurethane lacquer (litres)	—	—	—	—	5	—
Bituminous paint (litres)	—	10	11	11	—	—
Wallpaper paste	1 pkt (8 pts) per 4 rolls of wallpaper					

It should be stressed that the figures shown in the Table refer only to **average** covering capacities which are subject to fluctuation depending on the condition of the surface being treated, skill of the operative, quality of materials, etc.

Reference should be made wherever possible to manufacturers' catalogues and leaflets where more accurate information will be available.

7.10.3 Labour

The average outputs for skilled labour are shown in Table 7.3.

Table 7.3 Average skilled labour outputs in hours per 100 m^2

Operation	Walls		Ceilings
Prepare, 1 ct sealer, 2 cts emulsion paint	Conc.	24	26
	Brick	26	—
	Plast.	22	23
Prepare, 1 ct primer, 2 u/c, 1 f/c oil paint	Conc.	53	61
	Brick	54	—
	Plast.	52	60
Plastic compound (stippled finish)	Conc.	—	31
	Plast.	29	30
	Gen. wood surf.		*Glazed w/ws & doors (wood)*
Prepare, KPS 2 u/c, 1 f/c oil paint	59		Small panes 67
			Med. panes 64
			Large panes 61
			OE of casement 7½ hrs per 100 m
	Gen. wood surf.		*Glazed w/ws & doors (wood)*
T/up primer, 1 u/c, 1 f/c oil paint	38		Small panes 44
			Med. panes 42
			Large panes 40
			OE of casement 5 hrs per 100 m
	Gen. wood surf.		*Glazed w/ws & doors (wood)*
KPS only on wood surface before fixing	16		—
Prepare, 2 cts polyurethane varnish	21		Small panes 24
			Med. panes 23
			Large panes 22
	Gen. metal surf.		*Glazed w/ws & doors (metal)*
Prepare, 1 ct red lead primer, 2 u/c, 1 f/c	62		Small panes 83
			Med. panes 80
			Large panes 76
Prepare & prime only	20		Small panes 24
			Med. panes 23
			Large panes 21
	Rendered/conc. walls		
Prepare, 2 cts cement-based paint	29		
Prepare and 1 ct bituminous paint	15		
Prepare, size and hang wallpaper	2 hrs per roll (6 m^2)		

Work in narrow widths

The outputs in Table 7.3 are based on work over 300 mm wide and measured in square metres. However, where work is executed in narrow widths (i.e. not exceeding 300 mm wide) on frames, skirtings, pipes, etc., the unit of measurement is the metre and the width in stages of 150 mm is given in the bill description.

Due to the restriction in working conditions when painting in narrow widths, the labour outputs per square metre should be increased as follows before conversion to cost per metre:

n.e. 150 mm girth + 50 per cent
150–300 mm girth + 33⅓ per cent

Similarly, the cost per square metre for materials must be converted to cost per metre; however, no adjustment needs to be made in covering capacity. Where work is described as being within the ranges as given above, e.g. 0–150 mm, an **average** girth would be assumed, i.e. 75 mm.

Use and waste of brushes

An allowance of approximately 5 per cent of the total labour cost for any painting operation should be included in the rate to cover for the replacement of worn brushes, rollers and other sundry items of equipment used by the painters. The cost of providing dustsheets, ladders and scaffolding would normally be priced as a lump sum and included in the Preliminaries.

Work to ceilings and beams over 3.5 m high and work to staircase areas

Due to the difficulty of access when working on high ceilings and in staircase areas, labour outputs for all types of painting and decoration work carried out under these conditions should be increased by approximately 50 per cent.

External work

Allow an extra 10 per cent to the cost of labour and materials for all external work in recognition of additional preparation required, protection and additional non-productive time for adverse weather conditions.

Waste

The replacement of brushes, etc. is dealt with as explained above; however, an allowance for wastage of, say, 5 per cent should be included in the material cost which takes into account spillages, paint left over on completion of the work, etc.

The typical bill items to be priced in this section are as follows:

				£	p
A	Prepare, 1 coat sealer, 2 coats emulsion paint on plastered walls	1,210	m^2		
B	As A, but on plastered ceilings 3.5–5 m high	300	m^2		
C	Prepare, knot, prime and stop, 2 undercoats 1 finishing coat oil paint on general wood surfaces over 300 wide (internally)	172	m^2		
D	As C, but not exceeding 150 mm girth (internally)	330	m		
E	Touch-up primer, 1 undercoat, 1 finishing coat on wood window in medium panes (externally)	215	m^2		
F	Prepare metal surface, prime, 1 undercoat, 1 finishing coat on pipe not exceeding 150 mm girth	100	m		
G	Textured plastic coating to plasterboard ceiling over 300 wide in staircase areas	975	m^2		
H	Prepare, size and hang woodchip paper to plastered walls	600	m^2		

The build-up of unit rates for painting and decorating work

Item A – Emulsion paint to plastered walls

Basic prices:

Emulsion paint £8.00 per 5 litres
All-in rate for craftsman £4.50 per hour

Method

(a) *Materials (per 100 m^2):*

1 coat sealer	$1 \times 3\frac{3}{4}$ l	$= 3\frac{3}{4}$	
2 full coats	$2 \times 7\frac{1}{2}$ l	$= 15$	
		$18\frac{3}{4}$ l	
$18\frac{3}{4}$ l at £8.00 per 5 l			= £30.00
Add 5% waste			= 1.50
			£31.50 . . . [1]

(b) *Labour (per 100 m²)*:
Assume skilled operative can prepare surface, apply 1 coat of sealer and 2 coats of emulsion paint to plastered walls in 22 hrs

22 hrs at £4.50 per hour	= £99.00	
Add 5% use and waste of brushes	= 4.95	
	= £103.95	. . . [2]

Total rate ([1] + [2]) = £135.45 per 100 m²

Therefore total rate = $\frac{£135.45}{100}$ = £1.36 per m²

Item B – As Item A, but to plastered ceilings over 3.5 m not exceeding 5 m high

Method

(a) *Materials (per 100 m²)*:
Material cost as before at £31.50 . . . [1]

(b) *Labour (per 100 m²)*:
Assume skilled operative can prepare surface, apply 1 coat of sealer and 2 coats of emulsion paint to plastered ceilings in 23 hrs

23 hrs at £4.50 per hour	= £103.50	
Add 50% for work to ceilings 3.5–5 m high	= 51.75	
	£155.25	
Add 5% use and waste of brushes	= 7.76	
	£163.01	. . . [2]

Total rate ([1] + [2]) = £194.51 per 100 m²

Therefore total rate = $\frac{£194.51}{100}$ = £1.95 per m²

Item C – Prepare, KPS, 2 undercoats, 1 finishing coat oil paint on general wood surfaces

Basic prices:

Glasspaper 10p per sheet
Putty 40p per kg
Knotting £4.00 per litre
Wood primer £12.00 per 5 litres
Undercoat £11.00 per 5 litres
Gloss £11.00 per 5 litres

Method

(a) *Materials (per 100 m²):*

	(£)
8 sheets glasspaper at 10p sheet	= 0.80
2½ kg putty at 40p per sheet	= 1.00
¾ l knotting at £4.00 per l	= 3.00
10½ l wood primer at £12.00 per 5 l	= 25.20
2 × 9 l undercoat at £11.00 per 5 l	= 39.60
9 l gloss at £11.00 per 5 l	= 19.80
	£89.40
Add 5% waste	= 4.47
	£93.87 . . . [1]

(b) *Labour (per 100 m²):*

Assume skilled operative can prepare, KPS, and apply 2 undercoats and 1 finishing coat to general wood surfaces in 58 hrs

58 hrs × £4.50 per hour	= £261.00
Add 5% use and waste of brushes	= 13.50
	£274.50 . . . [2]

Total rate per 100 m² ([1] + [2]) = £368.37

Therefore total rate $= \frac{£368.37}{100} =$ £3.68 per m²

Item D – As Item C, but not exceeding 150 mm girth

Method

(a) *Materials*

Material cost as before at £93.87 per 100 m²

Therefore cost per m, assuming an average 75 mm girth

$= \frac{£93.87}{100} \times 0.075 =$ 7p per m . . . [1]

(b) *Labour*:

Labour cost as before at £274.50 per 100 m²

Add 50% for work in narrow widths = £137.25

Therefore cost per m, assuming an average 75 mm girth

$= \frac{£411.75}{100} \times 0.075 =$ 31p per m . . . [2]

Total rate ([1] + [2]) = 38p per m

Item E – Touch-up primer, 1 undercoat, 1 finishing coat on wood window in medium panes (externally)

Assumptions

1 Allow, say, 20 per cent of the area to be 'touched up'.

Method

(a) *Materials (per 100 m²):*

8 sheets glosspaper at 10p per sheet	= £0.80	
$10\frac{1}{2}$ l × 20 per cent primer at £12.00 per 5 l	= 5.04	
9 l undercoat at £11.00 per 5 l	= 19.80	
9 l glass at £11.00 per 5 l	= 19.80	
	= £45.44	
Add 5% waste	= 2.27	
	£47.71	. . . [1]

(b) *Labour (per 100 m²):*

Assume skilled operative can touch-up primer, apply 1 undercoat and 1 finishing coat to wood windows in 42 hrs

42 hrs at £4.50 per hour	= £189.00	
Add 5% use and waste of brushes	= 9.45	
	£198.45	. . . [2]

Total rate ([1] + [2]) = £246.16 per 100 m²

Add 10% for external work = £24.62

270.78

$$\text{Total rate} = \frac{£270.78}{100} = \underline{£2.71 \text{ per m}^2}$$

Note:

For windows in small panes allow 10 per cent extra paint
For windows in large panes allow 10 per cent less paint

Item F – Prepare, prime, 1 undercoat, 1 finishing coat on pipe not exceeding 150 mm girth

Assumptions

1 Assume an average girth of 75 mm (equal to 25 mm dia. pipe)

Basic prices:

Metal primer £17.00 per 5 litres

Method

(a) *Materials (per 100 m²):*

8 sheets glasspaper at 10p per sheet	= £0.80
10 l metal primer at £17.00 per 5 l	= 34.00
9 l undercoat at £11.00 per 5 l	= 19.80
9 l gloss at £11.00 per 5 l	= 19.80
	= 74.40
Add 5% waste	= 3.72
	£78.12 . . . [1]

(b) *Labour (per 100 m²):*
Assume skilled operative can prepare, apply 1 coat of primer, 2 undercoats and 1 finishing coat on general metal surfaces in 62 hrs

Deduct 1 undercoat at 14 hrs per 100 m²	
= 48 hrs at £4.50 per hour	= £216.00
Add 50% for work n.e. 150 mm girth	= 108.00
	£324.00
Add 5% use and waste of brushes	= 16.20
	£340.20 . . . [2]

Total rate per 100 m² ([1] + [2]) = £418.32

$$\text{Therefore rate per m} = \frac{£418.32}{100} \times 0.075 = \underline{31p}$$

Item G – Textured plastic coating to plasterboard ceilings in staircase areas

Basic prices:
Plastic compound £8.00 per 25 kg

Method

(a) *Materials (per 100 m²):*

40 kg plastic compound at £8 per 25 kg	= £12.80
Add 5% waste	= 0.64
	£13.44 . . . [1]

(b) *Labour (per 100 m²)*:

Assume skilled operative can apply plastic compound to plasterboard ceiling in 30 hrs

30 hrs at £4.50 per hour	=	£135.00
Add 5% use and waste of brushes	=	6.75
		£141.75
Add 50% for working in staircase areas	=	70.88
		£212.63 . . . [2]

Total rate per 100 m² ([1] + [2]) = £226.07

Therefore rate per m² $= \frac{£226.07}{100} = \underline{£2.26}$

Item H – Prepare, size and hang woodchip paper to walls

Basic prices:

Woodchip paper 85p per roll (10 m × 520 mm)
Wallpaper paste (8 pt pack) £1.20 each

Method

(a) Materials:

		£
1 roll woodchip paper	=	0.85
¼ pkt wallpaper paste	=	0.30
Size, stopping, say	=	0.15
		£1.30
Add 5% waste	=	0.07
		£1.37 [1]

(b) Labour:

Assume skilled operative can rub down, stop and size walls and hang woodchip paper at 1¾ hrs per roll

1¾ hrs at £4.50 per hour = £7.88 . . . [2]

Total rate per roll ([1] + [2]) = £9.25

Area per roll = 10 m × 520 mm = 5.2 m²

Therefore rate per m² $= \frac{£9.25}{5.2} = \underline{£1.78}$

7.11 Drainage

The drainage section of the Standard Method of Measurement provides for the measurement and pricing of the two main categories of work in drainage operations as follows.

1 *Pipe trenches*. Covering excavation work and the provision of pipework together with any beds, benchings and coverings which may be specified.
2 *Manholes, soakaways, cesspits, etc*. Covering the excavation and construction of the item.

Other sundry items include connections to sewers, (given as a provisional sum): the testing of the system, and the protection of the work measured as items and to be priced as lump sums where appropriate.

7.11.1 Pipe trenches

Excavation

The excavation of pipe trenches involves an operation similar to that of foundation trenches discussed earlier in section 7.1. However, there are significant differences in the measurement and documentation of the work, namely:

1 The excavation work is measured and priced in metres and not cubic metres with the depth to the nearest 250 mm being given in the bill description (Clause W.3.1).

   The width of the trench, whilst not being specifically stated, may be determined by the estimator since an indication of the size of the pipe to be laid in the trench is included in the bill decription (Clause W.3.3).
2 Unlike foundation trench excavation, the excavation of pipe trenches is deemed to include a number of operations in accordance with Clause W.3.4. which in the excavation and earthwork section are measured separately, i.e.

   (a) Earthwork support.
   (b) Treating bottoms of excavations.
   (c) Backfilling and compacting.
   (d) Disposal of surplus spoil.

   The cost of these operations must be established and expressed in terms of cost per metre before being added to the cost of excavation (per metre) in order to produce the total bill rate.

The principles involved in the calculation of the costs of the above activities are the same as those discussed earlier under the heading of excavation and earthwork.

A guide to trench widths required to accommodate pipework is given below.

	Diameter of pipe *100 mm* *(mm)*	*150 mm* *(mm)*	*225 mm* *(mm)*	*300 mm* *(mm)*
Trench depth:				
n.e. 1 m	600	700	800	900
1–2 m	750	850	950	1,050
over 2 m	900	1,000	1,100	over 1,200

Excavation below ground water-level, the breaking up of any concrete or brickwork, etc. encountered in the excavation, and the disposal of water are all measured in accordance with the rules for the excavation and earthwork section and priced accordingly.

Beds, benchings and coverings

Where beds, benchings and coverings are required, the unit of measurement is the metre with the width, thickness and type of material to be used being clearly stated together with the size of pipe to be laid.

Whether bedding materials are required or not will depend upon the strength of the pipe being used, the depth of the trench, nature of the ground, and the nature and extent of traffic expected on the surface. Concrete, sand or some form of granular fill such as pea gravel are the most common materials used for this purpose, the cost per tonne or per cubic metre being converted to cost per metre for inclusion in the unit rate.

The quantities of bedding/covering materials required in cubic metres per metre of trench are given below.

Type of bed/covering	*Diameter of pipe* *100 mm*	*150 mm*	*225 mm*	*300 mm*
100 mm bed and haunch	0.06	0.06	0.09	0.11
150 mm bed and haunch	0.08	0.09	0.12	0.14
100 mm bed and surround	0.13	0.16	0.21	0.27
150 mm bed and surround	0.15	0.18	0.23	0.30
100 mm bed	0.04	0.05	0.06	0.07
150 mm bed	0.07	0.08	0.09	0.10

Pipework

Pipework is priced in metres, the quantity being measured over all pipe fittings with such fittings described and enumerated as 'extra over' the length of pipe in which they occur. The principle involved in the pricing of these extra over items is therefore similar to that considered earlier in the plumbing work section, i.e. the cost of providing and fixing the fitting is calculated, from which must be deducted the cost of providing and fixing the corresponding length of pipe which is displaced by the fitting. Bends and taper pipes will displace approximately $\frac{1}{3}$ m of straight pipe whilst a junction will displace approximately $\frac{2}{3}$ m.

Pipes may be rigid (e.g. concrete, grey iron, vitrified clay, etc.) or flexible (e.g. steel and ductile iron, upvc, pitch fibre, etc.). The rigid pipes, although provided with flexible joints, will fail before significant deformation occurs whereas flexible pipes will withstand a considerable amount of deflection before collapse. The nature of the pipe to be used will be clearly specified as will its nominal size and method of jointing, all of which will influence the total cost.

Type of pipes	*Availability*	*Method of jointing*
Pitch fibre (plain or socketed ends)	75 to 100 mm dia. in lengths 2.45 m, 3.05 m	Ring coupling for plain ended pipes; rubber for sockets
upvc (plain or socketed ends)	110 to 160 mm dia. in 3 m, 6 m lengths	Polypropylene coupling and solvent-welded for plain ends; 'O' ring for sockets
Vitrified clay (plain or socketed ends)	100 to 400 mm dia. in 1.6 m lengths	Polypropylene coupling with sealing ring; cement mortar (1 : 3)
Concrete (ogee joints)	150 to 600 mm dia., 2 m lengths	Cement mortar (1 : 3)
Spun iron (socketed joints)	75 to 225 mm dia., 2 m, 3 m lengths	Spun yarn and molten lead, caulked flush

Waste. The extent of the waste allowance needed for pipework will depend largely on the complexity of the layout and the number of short branches required which could necessitate a considerable amount of cutting. In such cases, since short cut lengths of pipe have to be used, the ratio of jointing couplings to pipe is therefore greater and the cost per metre of pipework correspondingly higher. However, where long uninterrupted lengths of drainage are concerned, virtually no cutting will be needed and the waste factor can be limited to an allowance for breakages.

Average waste allowances may be in the order of 5 per cent for clay and concrete pipes and fittings, and $2\frac{1}{2}$ per cent for upvc, pitch fibre and spun iron items.

7.11 2 Manholes, soakaways, etc.

The measurement and documentation of the work involved in the provision of manholes and the like follows the rules contained in the respective work sections of excavation, (7.1) concrete work (7.2) and brickwork (7.3) considered earlier. Special mention should be made, however, of the method of measuring and pricing of concrete benching and channels, etc. which are to be enumerated separately stating the type, quality and size of the item (see Clause W.7.5).

7.11.3 Guide to average labour and mechanical plant outputs

Trenches

Normally, a wheeled or tracked excavator with backactor of bucket capacity, say ¼ m³, would be used for the excavation of trenches together with a labourer acting as a banksman. Consideration must be given, however, to the possibility of having to carry out a certain amount of hand digging in situations where access for the machine is difficult or impossible. The site visit and examination of the layout drawing should provide the estimator with sufficient information to enable him to determine the approximate ratio of machine digging to hand digging, whereupon he would be able to calculate a single composite rate for the excavation based upon the appropriate mix of the two methods.

Method of excavation	*Output in m³ per hour in ordinary ground*			
	Depth of trench			
	Up to 1 m	*1–2 m*	*2–3 m*	*3–4 m*
Mechanical excavator (¼ m³ bucket capacity and banksman)	6	5½	5	4½
Hand excavation	0.43	0.37	0.25	0.20

Notes: (a) The cost per m³ must be converted to cost per m.
(b) Adjustments to the above outputs should be made for excavation in different types of ground – see Excavation work section (7.1) notes on 'multipliers'.

(a) *Backfilling and removing surplus spoil.* Backfilling operations may have to be carried out entirely by hand where stones and large lumps of clay, etc. have to be removed. Where this is the case, it will take approximately 1 hr per m³ for this operation which would include for compacting in layers as specified.

On average, approximately three-quarters of the total excavated material will be backfilled into the trench with the surplus being either spread and levelled on the site or removed to some suitable tip. These disposal items would be priced as indicated in the excavation work section.

(b) Earthwork support and treating bottoms of excavation. These items would be priced in accordance with the principles stated in the excavation work section, remembering to convert from cost per m^2 to cost per m.

Beds, benchings and coverings	*output-unskilled labour*
Granular materials for beds and coverings to pipes	2 hrs per m^3
Concrete for beds and coverings (based on site-mixed concrete – reduce by half for ready-mix)	3 hrs per m^3

Pipework, etc.

The NWRA provides for a 'pipelayer and jointer' to be paid a few pence extra per hour above the rate for a labourer in recognition of the extra skill and responsibility involved.

Pipelayer outputs	*100 mm dia.*	*150 mm dia.*
1.6 m length vitrified clay	0.10 hr/pipe	0.15 hr/pipe
pipe joint	0.15 hr each	0.20 hr each
2.45 m length pitch fibre	0.20 hr/pipe	—
pipe joint	0.10 hr each	—
3 m length upvc pipe joint	0.15 hr/pipe	0.20 hr/pipe
	0.10 hr each	0.15 hr each
2 m length concrete pipe	—	0.03 hr/pipe
joint (ct. and sand)	—	0.25 hr each
2 m length spun iron pipe	0.50 hr/pipe	0.60 hr/pipe
joint (caulked lead) (plumber and mate)	0.50 hr each	0.60 hr each

Bends – Allow time for laying ½ m of pipe plus 1 joint
Junctions – Allow time for laying ¾ m of pipe plus 2 joints

Notes: (a) For vertical pipework, add 100 per cent to labour outputs.
(b) For branches n.e. 3 m in length, add 50 per cent to labour outputs and an additional 10 per cent waste on pipes due to extra cutting.

Sundries	
100 mm dia. stoneware gulley	0.50 hr each
Road gulley	1.00 hr each
Build in end of 100 mm dia. pipe to 1B thick wall	0.15 hr each
Medium duty manhole cover and frame	2.00 hrs each
Step iron	0.20 hr each

The typical bill items to be priced in this section are as follows:

					£	p
A	Excavate drain trench for pipe not exceeding 200 dia. not exceeding 2 m deep, including earthwork support, grading bottoms, backfilling and removing surplus from site	400	m			
B	Granular filling of pea gravel in 150 mm bed and surround to 100 mm dia. pipe	400	m			
C	Concrete mix (1 : 3 : 6) in 150 mm bed and haunch to 150 mm dia. pipe	150	m			
D	100 mm dia. vitrified clay drain pipes with sleeve joints laid horizontally in trench	400	m			
E	As D, in branches not exceeding 3 m	75	m			
F	Extra for 100 mm equal junction	30	no.			
G	75 mm dia. pitch fibre conduit	272	m			
H	100 mm dia. vitrified clay gulley set on and surrounded with 150 mm concrete mix (1 : 3 : 6) and joint to drain	15	no.			
I	610 × 457 grade B medium-duty cast iron manhole cover and frame	6	no.			

The build-up of unit rates for drainage work

Item A – Excavate drain trench for pipe n.e. 200 dia. n.e. 2 m deep, average 1.25 m deep including earthwork support, grading bottom, backfill and removing surplus

Assumptions:

1 The ground is 'ordinary ground'.
2 The excavation work is carried out by $\frac{1}{4}$ m^3 bucket excavator and the backfilling and compacting by hand.

Basic prices: All-in rate for labour £4.00 per hour
All-in rate for $\frac{1}{4}$ m^3 excavator and driver (including fuel, oil, etc.) £12.00 per hour

Method

(a) *Excavation*

Hourly cost of excavator and banksman
= £12.00 + £4.00 (+ 10p NWRA) = £16.10

Assume output of excavator is $5\frac{1}{2}$ m^3 per hour

Therefore cost per m^3 = $\frac{£16.10}{5\frac{1}{2}}$ = £2.93

Quantity of excavation per m of trench 750 mm wide and 1.25 m deep = 0.94 m^3
Therefore cost of excavation per m.
= £2.93 × 0.94 = £2.75 . . . [1]

(b) *Grading and ramming bottom of trench*:

Assume labourer can grade and ram 20 m of trench bottom 750 mm wide in 1 hr

Therefore cost per m = $\frac{£4.00}{20}$ = 20p . . . [2]

(c) *Earthwork support*:

(assuming earthwork support is required)
Amount of earthwork support per m of trench
= 2× 1.00 × 1.25 =$2\frac{1}{2}$ m^2
Using rate calculated earlier in 'Excavation and earthwork' section of £2.00 per m^2
Therefore cost per m = $2\frac{1}{2}$ × £2.00 = £5.00 . . . [3]

(d) *Backfilling and compacting*:

Assume labourer can backfill 1 m^3 in 1 hr
Therefore cost per m^3 = £4.00
Assume $\frac{3}{4}$ of the excavated material is backfilled
Therefore quantity of backfilling per m of trench
= $\frac{3}{4}$ × 0.94 m^3 = 0.71 m^3
Therefore cost per m = 0.71 × £4.00 = £2.84 . . . [4]

(e) *Removing surplus material from site*:

Quantity of surplus per m of trench = 0.94–0.71 m^3
= 0.23 m^3

Using rate calculated earlier in 'Excavation and Earthwork' section of £3.67 per m^3
Therefore cost per m =0.23 × £3.67 = 84p . . . [5]

Total rate ([1] + [2] + [3] + [4] + [5]) = £11.63 per m

Item B – Granular filling of pea gravel in 150 mm bed and surround to 100 mm dia. pipe

Basic prices:

Pea gravel £15.00 per tonne

Method

(a) *Materials*:

Cost of pea gravel delivered to site per tonne	= £15.00
Add 5% waste	= 0.75
	£15.75

Pea gravel weighs approx 1.60 tonnes per m^3
Therefore cost per m^3 = 1.60 × £15.75 = £25.20
Quantity of material required per m of trench = 0.15 m^3
Therefore cost per m = 0.15 × £25.20 = £3.78 . . . [1]

(b) *Labour*:

Assume labourer can fill pea gravel in trench to form bed and surround at 2 hrs per m^3
Therefore cost per m = 2 × £4.00 × 0.15 = £1.20 . . . [2]

Total rate ([1] + [2]) = £4.98 per m

Item C – Concrete mix (1 : 3 : 6) in 150 mm bed and haunch to 150 mm dia. pipe

Assumptions

1 Concrete is to be obtained ready-mixed delivered to site in trucks.

Basic prices:

(1 : 3 : 6) mix ready-mixed concrete £50.00 per m^3

Method

(a) *Materials*:

Cost of ready-mixed concrete delivered to site per m^3	= £50.00
Add 2½% waste	= 1.25
	£51.25

Quantity of material required per m of trench = 0.09 m^3
Therefore cost per m = 0.09 × £51.25 = £4.61 . . .[1]

(b) *Labour*:

Assume labourer can spread concrete in bed and haunch to 150 mm dia. pipe at 1½ hrs per m^3
Therefore cost per m = £4.00 × 1½ × 0.09 = 54p . . .[2]

Total rate ([1] + [2]) = £5.15 per m

Item D – 100 mm dia. vitrified clay drain pipes with sleeve joints laid and jointed horizontally in trench

Basic prices:

1.6 m length 100 mm dia. vitrified clay pipe £3.00 each
100 mm dia. couplings £1.20 each
All-in rate for pipelayer and jointer £4.00 + 12p (NWRA)

Method

(a) *Materials*:

Cost of 1.6 m length of 100 dia. pipe delivered to site	= £3.00	
1 coupling per length at £1.20 each	= 1.20	
Allow for lubricant, say	= 0.04	
	£4.24	
Add 5% unloading and waste	= 0.21	
	£4.45	. . . [1]

(b) *Labour*:

Assume pipelayer and jointer can lay 1.6 m length of pipe and make 1no. joint in 0.10 + 0.15 hrs
Therefore cost per length = £4.12 × 0.25 = £1.03 . . .[2]

Therefore total rate per 1.6 m length ([1] + [2]) = £5.48

Therefore rate per m = $\frac{£5.48}{1.6}$ = £3.43

Item E – As Item D, but in branches not exceeding 3 m

Method

(a) *Materials*:

Material cost as before + 10% additional waste for runs n.e. 3 m long
= £4.45 + 0.45 = £4.90 . . . [1]

(b) *Labour*:

Labour cost as before + 50% for additional cutting and fitting in branches
= £1.03 + 0.52 = £1.55 . . . [2]

Total rate per 1.6 m length ([1]+ [2]) = £6.45

Therefore rate per m = $\frac{£6.45}{1.6}$ = £4.03

Item F – Extra for 100 mm dia. equal junction

Basic prices:

100 × 100 mm vitrified clay junction £4.50 each

Method

(a) *Materials*:

Cost of junction delivered to site	= £4.50 each
2no. couplings at £1.20 each	= 2.40
Lubricant, say	= 0.08
	£6.98
Add 5% unloading and waste	= 0.35
	£7.33 . . . [1]

(b) *Labour*:

Allow time for fixing $\frac{3}{4}$ m of pipe = 0.10 hr × 0.75 = 0.075
Allow time for making 2no. joints at 0.15 hr each = 0.30
= 0.375 hr
Therefore cost of fixing junction = 0.375 × £4.12
= £1.55 . . . [2]

Total rate ([1] + [2]) = £8.88 each

Since fittings such as junctions, bends, etc. are measured **extra over**, the cost of the length of drain pipe displaced by the junction must be deducted
Deduct:
$\frac{2}{3}$ m pipe × £3.43 (from Item D)
= £2.29
Therefore net rate for junction = £8.88–£2.29
= <u>£6.59 each</u>

Item G – 75 mm dia. pitch fibre conduit

Basic prices:

75 mm dia. pitch fibre pipe £6.30 per 2.45 m length
Ring couplings 75 mm dia. £1.50 each

Method

(a) *Materials*:

Cost of 2.45 m length of 75 mm dia. pitch fibre pipe	= £6.30
1no. coupling at £1.50 each	= 1.50
Allow for lubricant, say	= 0.03
	£7.83 . . . [1]

(b) *Labour*:
Assume pipelayer and jointer can lay 2.45 m length of 75 dia. pitch fibre pipe and make 1no. joint
in 0.20 + 0.10 hrs = 0.30 hr
Therefore cost per length = 0.30 × £4.12 = £1.24 . . . [2]

Total rate ([1] + [2]) = £9.07 per length

Therefore rate per m = $\frac{£9.07}{2.45}$ = £3.70

Item H – 100 mm dia vitrified clay gulley set on and surrounded with 150 mm concrete mix (1 : 3 : 6) and joint to drain

Basic prices:
100 mm dia. clay 'P' trap gulley £9.50 each

Method

(a) *Materials*:

Cost of 100 mm dia. gulley delivered to site	= £9.50 each
1no. 100 dia. coupling at £1.20 each	= 1.20
Allow for lubricant, say	= 0.04
Allow for concrete in bed and surround 0.05 m³ at £50.00 per m³	= 2.50
	£13.24
Add 5 per cent unloading and waste	= 0.66
	£13.90 . . . [1]

(b) *Labour*:
Assume pipelayer and jointer can fix, joint and bed, gulley in position, including surrounding in concrete, in ½ hr
Therefore cost = £4.12 × 0.5 = £2.06 . . . [2]

Total rate ([1] + [2]) = £15.96 each

Item I – Grade B medium-duty manhole cover and frame 610 × 457 mm

Basic prices:
610 × 457 mm grade B medium-duty c.i.
manhole cover and frame £65.00 each

Method

(a) *Materials*:

Cost of manhole cover and frame delivered to site	= £65.00	
Add 5% unloading and waste	= 3.25	
Allow for cement mortar for bedding frame, say	= 0.20	
Allow for grease for bedding cover, say	= 0.10	
	£68.55	. . . [1]

(b) *Labour*:

Assume skilled operative can fix c.i. manhole cover and frame in 2 hrs

Therefore cost = £4.50 × 2 hrs = £9.00 . . . [2]

Total rate ([1] + [2]) = £77.55 each

Chapter 8

Methods of approximate estimating

Before a prospective client can give approval for detailed and costly design work to proceed, he must first be sure that his funds are sufficient to cover the full development costs, including the legal and professional fees, involved in the provision of the building. If at this early stage in the process, the development appraisal proves the building to be too costly, then amendments to the scheme can be made or the project abandoned with as little wasted effort and cost as possible.

The early preparation of an approximate estimate of cost is therefore essential in order for the client to plan his next move. The task of preparing this approximate estimate often falls upon the private quantity surveyor who, often with very little information to work on, is required to produce this crucial figure as accurately as possible. Having been appointed to the job of preparing an approximate estimate of the cost of the building, the quantity surveyor can avail himself of one of the following methods in common use:

Cost per functional unit
Cost per square metre of floor area.
Approximate quantities.

8.1 Cost per functional unit

This method is used mainly for public sector works where large sums of money are involved but where little, if any, serious design work has been carried out at this stage and must be regarded as a very rough guide suitable for preliminary investigations only. It involves the establishment of target costs for functional units provided, with such costs being frequently revised in the light of current Government and local authority expenditure plans.

Having determined the number of functional units required (i.e. hospital beds, car parking spaces, school places, etc.) the total cost of the project can then be gauged from multiplying the number of units provided by the cost of providing each unit. The cost of providing the unit itself would have to be calculated from information extracted from the documentation and pricing of previous similar contracts, with costs suitably adjusted for inflation, changes in design, state of the market, etc.

Whilst the calculation of the approximate estimate is very simple, the problems involved in establishing a realistic cost per functional unit are considerable since at this stage there are so many unknown factors to be taken into account, each of which demands arbitrary assumptions to be made in respect of their cost implications.

8.2 Square metre method

An improvement upon the cost per functional unit method is the square metre of floor area system whereby, providing some basic design work has been done, the total floor area of all storeys, measured between external walls and without deduction for internal walls, stairwells, etc., is calculated. This area is then multiplied by an appropriate rate per square metre of floor area for that type of building in order to obtain the total approximate cost.

Unfortunately this method suffers from the same problems as the cost per functional unit method in that the rate per square metre obtained from an analysis of previous similar projects is difficult to determine with any accuracy. The system fails to take into account plan shape, storey height, number of storeys and changes in specification and, again, must remain no more than a rough guide to cost. However, its merit lies in the fact that the calculation is quick and straightforward and that cost is expressed in terms of square metre of floor area which is probably more meaningful to the client with it being directly related to usable floor space.

8.3 Approximate quantities

The method which makes use of approximate quantities can be regarded as being the most accurate of all approximate estimating techniques. However, it relies on a significant amount of design work having been completed and can be a lengthy process since it is essentially a shortened version of the preparation and pricing of a traditional bill of quantities.

The most significant items of work with regard to cost are identified and measured in accordance with the guidelines indicated below, with less important items, labours and sundries being ignored, although their omission will influence the rates used for remaining items. The main items having been identified are then allocated to 'elements' appropriate to the scheme under consideration and an 'all-in' composite rate per unit of measurement calculated for the batch of operations comprising the provision of these elements.

Example

Element: Traditional strip foundations and cavity walling up to DPC

Operations involved	*Normal unit of measurement*	*Composite unit of measurement*
A. Foundation trench excavation	m^3	
B. Backfilling of excav. material	m^3	
C. Disposal of surplus spoil	m^3	
D. Earthwork support	m^2	
E. Level and compact	m^2	m
F. Concrete in foundations	m^3	
G. Brickwork in cavity wall	m^2	
H. Forming cavity	m^2	
I. Concrete in cavity fill	m^3	
J. DPC	m	
K. Adjustment for facings below ground	m^2	

The most appropriate composite unit of measurement for approximate estimating purposes for this particular element would be the metre which necessitates the cost of each of the above items to be expressed in terms of this unit; therefore the quantity and subsequently the cost of each item per metre of trench must be calculated for incorporation into the total rate. Having thus established a rate per metre, the total cost of the element can be found by multiplying this rate by the total quantity involved, i.e. in this case, the perimeter of the building measured in metres.

As before, an appropriate rate for each operation would be determined from an analysis of previous similar projects suitably adjusted as described for the current circumstances.

Whilst there are difficulties involved in assessing accurately the extent of the adjustment to be made, the margin of error is reduced when compared with the previous two methods since the cost of each item may be adjusted individually and as such rates more accurately reflect actual costs.

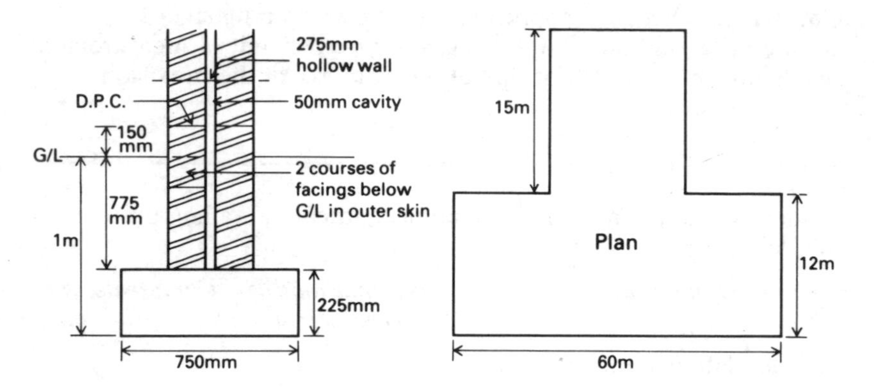

Operation	*Quantity per metre of ext. perimeter*		*Rate*	*Rate per metre of ext. perimeter* (£)
A. Found. tr. excav.	1.0 × 0.75 × 1.0 m	= 0.75 m³	6.00	4.50
B. Backfilling	2/0.775 × 0.225 × 1.0 m	= 0.35 m³	3.00	1.05
C. Disposal of surpl.	A − B	= 0.40 m³	5.00	2.00
D. Earth. support	2/1.0 × 1.0 m	= 2 m²	1.50	3.00
E. Level & compact	0.75 × 1.0 m	= 0.75 m²	0.50	0.38
F. Conc. in founds.	0.75 × 0.225 × 1.0 m	= 0.17 m²	50.00	8.50
G. Cmn. bks. in cav. wall	2/0.925 × 1.0 m	= 1.85 m²	20.00	37.00
H. Forming cavity	0.925 × 1.0 m	= 0.93 m²	1.20	1.12
I. Filling to holl. wall	0.775 × 0.05 × 1.0 m	= 0.04 m³	60.00	2.40
J. DPC	2/1.0 m	= 2 m	0.50	1.00
K. Adjust. for facings	ddt. 0.30 × 1.0 m cmns. add ditto in fcgs.	= 0.30 m²	(30.00 − 20.00)	3.00
TOTAL RATE PER METRE FOR WORK UP TO DPC				£63.95

From the example, it can be seen immediately which items are most significant and which therefore demand extra care and attention when calculating the corresponding rate. Having such information to hand in the early stages can help considerably in the design of the building and choice of materials to be used in order to achieve the most effective use of resources.

Total perimeter of building:	2 × 60 m =	120 m
	2 × 12 m =	24 m
	2 × 15 m =	30 m
		174 m

Total cost for work up to DPC = £63.95 × 174 m = £11,127

Other examples of suitable elements to be priced may be:

(a) *Ground floors* – unit of measurement the square metre and comprising: oversite excavation; disposal of spoil; level and compact; hardcore bed; level, blind and compact; concrete slab; mesh reinforcement; damp-proof membrane; cement and sand screed; floor finish (tiles, carpet, etc.).

(b) *Windows* – unit of measurement the square metre (or enumerated) comprising: window, frame and cill; bedding and pointing in mastic; glass: ironmongery; window board; lintel; plaster to reveals; angle beads; all decorations to window and opening; fair return on facings; closing cavity.

8.4 Sources of reference

The most reliable source of information in respect of rate determination would be the contractor's own records of cost prepared from earlier contracts. However, for the private quantity surveyor, this source of reference is unlikely to be available, in which case reliance must be placed upon previously priced bills of quantities, trade journals, pricing books and the Building Cost Information Service. This latter publication, produced by the RICS, provides up-to-date information relating to tender price and cost indices, forecasts, labour and daywork rates, material prices, regional variations in trends, etc., and is particularly useful for giving background information in respect of all aspects of the building process which have an influence on cost.

Chapter 9

Pro rata and analogous rates

One of the most common and yet one of the most important tasks to be undertaken by the quantity surveyor in post-contract practice is the measuring and valuing of variations carried out by the contractor following the issue of architect's instructions under the contract.

In order to establish some guidelines for the parties involved in such valuations, the JCT Standard Form of Building Contract provides a number of rules stating the procedure to be followed (Clause 13.5). In many instances, variation work, whether additions or omissions, will be priced using rates contained in the bill of quantities and should present few difficulties in arriving at a fair valuation. However, despite these guidelines, there will be numerous occasions where bill rates are not appropriate for valuing variations, such as when work is of a similar character to that contained in the bill of quantities but is executed under different conditions and/or where the quantity of the work changes significantly. In these circumstances bill rates, as far as possible, are to be used as a **basis** for determining a fair valuation, having due regard to the changed nature of the work.

There is therefore sufficient scope for the quantity surveyor to exercise his own personal judgement in assessing the amount by which an existing bill rate should be adjusted in order to produce a revised price which accurately reflects a fair reimbursement to the contractor for carrying out the work. In pursuance of this task it is essential for him to possess a thorough knowledge of the composition of unit rates, the method of pricing of each of the resources needed for that particular operation and to bear in mind the fact that bill unit rates do not always accurately represent the **cost** of carrying out work and may be distorted due to:

1 Profit and overheads priced elsewhere in the bill (e.g. as a lump sum in the Preliminaries).
2 An allowance for inflation on firm price contracts (alternatively, such allowances may have been included as a lump sum in the Preliminaries).
3 Deliberate under- or over-pricing of rates so as to influence cash flow at particular stages in the contract or in anticipation of future variations.
4 Mechanical plant being included in the rate, in the plant items at the beginning of the corresponding work section, or in the Preliminaries. (Without further examination of the bill of quantities and/or consultation with the contractor, it would be impossible to detect the estimator's method of pricing).
5 The arbitrary allocation of costs between a number of items in the bill which comprise one main activity priced initially as a lump sum by the estimator and which subsequently requires to be broken down into separate items on submission of the priced bill of quantities for checking.

Being aware of the pitfalls involved in the adjustment of bill rates, the following suggestions provide a basis upon which suitable calculations may be made:

Step 1. Establish the percentage addition allowed in the original rate in respect of overheads and profit (and any other factors) and deduct the corresponding amount from the rate in order to reduce it to net cost.

Step 2. Identify the resources involved in the operation and isolate the unknown factors by calculating that part of the rate which can be readily verified by way of invoices, quotations, plant hire rates, etc. (normally the unknown factor will be the labour element).

Step 3. Adjust the unknown factor by pro rata (i.e. in proportion) method where applicable or on some other suitable *ad-hoc* basis.

Step 4. Add back the relevant material and mechanical plant element (if applicable), the nature and cost of which again being suitably adjusted in the light of the changed circumstances.

Step 5. Add back the original percentage for overheads and profit to produce the new rate.

Examples of the use of these methods are given below

Example 1.
To determine the accuracy of an existing bill rate
'65 mm facing bricks in skin of hollow wall in stretcher bond in cement lime mortar (1 : 1 : 6) pointed with a neat flush joint as work proceeds.'

. . . 750 m^2 at £22.50 per m^2

Step 1. Assuming 15 per cent for overheads and profit, deduct this from the rate to obtain the net cost:
If x is the net cost before the addition of overheads and profit:

$$x + 0.15x = £22.50$$
$$x(1 + 0.15) = £22.50$$
$$x = \frac{£22.50}{1.15} = £19.57$$

Step 2. Resources needed: Bricks, mortar and labour – the cost of the bricks and mortar may be readily calculated from quotations and invoices.
Assume: All-in rate for : craftsman £4.50 per hr
: labourer £4.00 per hr
: 10/7 mixer £1.60 per hr

Building sand £11.00 per tonne
Cement £64.00 per tonne
Lime £72.00 per tonne

Cost of bricks

Cost of facings delivered to site per 1,000	=	£200.00
Unload and stack 1 hr at £4.00 per hr	=	4.00
		£204.00
Add 10% waste	=	20.40
		£224.40

No. of bricks required per m^2 = 59
Therefore cost of bricks per m^2

$$= \frac{£224.40 \times 59}{1,000} = \underline{£13.24}$$

Cost of mortar

Cost of cement delivered to site per tonne	=	£64.00
Unload and stack 1 hr at £4.00 per hour	=	4.00
		£68.00
Cost of lime delivered to site per tonne	=	£72.00
Unload and stack 1 hr at £4.00 per tonne	=	4.00
		£76.00

Cost of 1 : 1 : 6 mix

1 m^3 of cement costs £68.00 × 1.44	=	£97.92
1 m^3 of lime costs £76.00 × 0.6	=	45.60
6 m^3 of sand cost £11.00 × 1.52 × 6	=	100.32
	=	£243.84
Add 33⅓% shrinkage	=	81.28
		£325.12

Therefore cost per m^3 = $\frac{£325.12}{8}$ = £40.64

Cost of mixing

All-in rate for mixer	=	£1.60
Operator at £4.00 + 12p (NWRA)	=	4.12
		£5.72

Assume: 10/7 mixer produces 2 m^3 of mortar per hour cost per m^3

= $\frac{£5.72}{2}$ = £2.86

Therefore total cost of mixed mortar	=	£40.64 + 2.86
	=	43.50
Add 5% waste	=	2.18
	=	£45.68 per m^3

Quantity of mortar required per m^2 of brickwork half-brick thick = 0.018 m^3

Therefore cost of mortar per m^2

= £45.68 × 0.018

= 82p

Therefore amount allowed in the bill rate for unknown factor (labour):

= £19.57 – (£13.24 + 82p)

= £5.01

It should now be considered whether the £5.01 is adequate for the labour required to build 1 m^2 of brickwork.

Hourly gang cost based on 2 : 1 gang would be

£4.50 + £4.50 + £4.00 = £13.00

Therefore $\frac{£13.00 \times 59}{\text{No. of bricks laid by gang in 1 hour}}$ = £5.01

Therefore no. of bricks laid per hour = $\frac{£13.00 \times 59}{£5.01}$

= 153 or 77 bricks per hour per bricklayer

Conclusion

It would seem that the labour element has been underpriced if it is considered that an output of, say, 40 facing bricks per hour per man is 'reasonable'. The calculations show that the output required in this instance is approximately double that figure, and as such the labour cost of £5.01 is approximately half of what it should be.

Example 2

To build up a rate for A193 mesh reinforcement weighing 3.02 kg per m² in foundations 750 mm wide using as a basis an existing bill rate for A142 mesh weighing 2.22 kg per m² in ground slabs of £2.20 per m²

Step 1. Deduct for overheads and profit (say 15%)

$$\text{Therefore net cost} = \frac{£2.20}{1.15} = £1.91$$

Step 2. Resources needed: Unskilled labour, mesh reinforcement, sundries (tying wire, distance blocks).

Assume: 2.2 × 3.6 m sheet of A142 mesh costs £11.00.
Therefore material content of rate would be:

Cost of A142 mesh reinforcement delivered to site per sheet	=	£11.00
Allow 2 labourers 1/20 hr per sheet unloading at £4.00 per hour	=	0.40
		£11.40
Add 15% for laps and waste	=	1.71
		£13.11
Allow for tying wire, spacing blocks, say	=	0.50
		£13.62

$$\text{Therefore cost per m}^2 = \frac{£13.62}{(2.2 \times 3.6)} = \underline{£1.72}$$

Therefore amount allowed in bill rate for unknown factor (labour)

= £1.91 − 1.72
= 19p

Step 3. Adjust labour rate as follows:
The labour content will be greater since (a) A193 mesh is heavier and (b) it is laid in narrow widths in foundations instead of in full sheets in slabs which necessitates additional cutting. Assume the extent of the increase in labour content is, say, 100 per cent.

Therefore new labour cost = 19p × 100%
= 38p

Step 4. New material rate: Since the cost of A193 mesh can be verified from invoices, etc, the new material cost can be calculated as follows.

Assume: 2.2 × 3.6 m sheet of A193 mesh costs £14.00 per sheet.

Cost of A193 mesh reinforcement delivered to site per sheet	= £14.00
Unload and stack, say 10% longer due to heavier weight	= 0.44
	£14.44
Allow for extra waste due to cutting to 750 mm widths, say 20%	= 2.89
	£17.33
Allow for tying wire, spacing blocks, say	= 0.50
	£17.83

Therefore cost per m² = $\frac{£17.83}{(2.2 \times 3.6)}$ = £2.25

Therefore net cost per m² of new operation = £2.25 + 0.38
= £2.63

Step 5. Add back relevant percentage for overheads and profit (15 per cent)

= £2.63 × 15% = 0.39
New rate per m² = £2.63 + 0.39 = £3.02

The unit of measurement in the bill of quantities for mesh reinforcement in single widths in foundations and the like is the metre, therefore the rate per m² must be converted to a rate per m.

New rate per m = £3.02 × 0.750 m
= £2.27

Example 3
To calculate a revised bill rate following a change in specification
'Calculate the revised bill rate for 25 mm thick tongued and grooved softwood boarded flooring using 175 mm in lieu of 125 mm wide boards where the original bill rate is £16.00 per m^2.

Step 1. Deduct for overheads and profit as before.

$$\text{Therefore net cost} = \frac{£16.00}{1.15} = £13.91$$

Step 2. Resources needed: Skilled labour, floor boarding, steel floor brads.
Assume: 25 × 125 mm softwood t. & g. flooring costs £125.00 per 100 m
25 × 175 mm softwood t. & g. flooring costs £148.00 per 100 m
60 mm cut steel floor brads cost £3.00 per kg (approx. 250)
All-in rate for joiner £4.50 per hour

Cost of flooring using 125 mm wide boards

Cost of boarding delivered to site per 100 m	= £125.00
Unload and stack ¼ hr at £4.00 per hour	1.00
	£126.00
Floor brads – allow 5 per m 500 per 100 m (2 kg)	= 6.00
	£132.00
Add, say, 5% waste	6.60 per 100 m
	£138.60

$$\text{Quantity required per m}^2 \text{ of flooring} = \frac{1\ \text{m}^2}{(125 - 10)}$$

$$= \frac{1\ \text{m}^2}{115\ \text{mm}} = \underline{8.7\ \text{m}}$$

$$\text{Therefore cost of materials per m}^2 = \frac{£138.60 \times 8.7\ \text{m}}{100\ \text{m}}$$

$$= \underline{£12.06}$$

Therefore amount allowed in bill rate for unknown factor (labour)

$$= £13.91 - £12.06$$
$$= \underline{£1.85 \text{ per m}^2}$$

Step 3. Adjust labour rate as follows:

(a) Adjust for revised length of flooring needed to provide 1 m^2 using 175 mm wide boards.

Quantity of 175 mm wide boards needed per m^2 of flooring $= \frac{1\ m^2}{(175 - 10)}$

$= \frac{1\ m^2}{165\ mm}$

$= \underline{6.06\ m}$

Therefore new labour rate $= \frac{£1.85 \times 6.06}{8.7}$

$= £1.29$

(b) Adjust for heavier boards – say 10 per cent extra for handling and fixing

$= £1.29 \times 1.10$
$= \underline{£1.42 \text{ per } m^2}$

Step 4. New material rate.

Cost of boarding (175 mm wide boards) delivered to site per 100 m	= £148.00
Unload and stack $\frac{1}{3}$ hr at £4,00 per hour	= 1.33
	£149.33
Allow 2 kg of floor brads as before	= 6.00
	£155.33
Add, say, 5% waste	= 7.77
	£163.10

Therefore cost of materials per $m^2 = \frac{£163.10 \times 6.06\ m}{100\ m}$

$= \underline{£9.88}$

Therefore net cost of new operation = £1.42 + £9.88

= £11.30 per m^2

Step 5. Add back relevant percentage for overheads and profit:

= £11.30 × 15% = £1.70

New rate per m^2 = £11.30 + 1.70 = £13.00

Example 4.
To calculate the cost implications of changes in contractual conditions

The following bill rates for preliminary items as described were established for a 12-month contract for the construction of a supermarket:

(a)	Provision of site office	£2,500.00
(b)	Temporary telephone	£2,000.00
(c)	Water for the works	£1,500.00
(d)	Non-productive overtime	£3,000.00

Calculate revised bill rates for the above in respect of a 6-week reduction in the contract period.

(a) Provision of site office

Step 1. Deduct for overheads and profit, say, 15 per cent.

$$\text{Therefore net cost} = \frac{£2{,}500.00}{1.15} = £2{,}174$$

Step 2. Isolate the fixed costs from the time-related costs:

Fixed costs: delivery, handling, erecting and dismantling, sundry repairs.

Time related costs: hire rate, regular cleaning and maintenance.

Assume fixed costs as follows:

Loading at yard – 4 labourers for 1 hr at £4.00 per hour	–	£16.00
1 hr standing time for lorry and driver at £12.00 per hour	=	12.00
Journey time to site and returning to yard 2 hrs at £12.00	=	24.00
Unloading at site – 4 labourers for ½ hr at £4.00 per hour	=	8.00
Erecting site office – 2 joiners and 2 labourers for 4 hrs (2 × £4.50 + 2 × £4.00) ×4 hrs	=	68.00
	=	£128.00
Allow same again for dismantling and returning to yard on completion of contract	=	128.00
		£256.00
Allow, say, £50.00 for sundry repairs	=	50.00
		£306.00

Therefore time-related costs = £2,174.00 – £306.00
= £1,868.00

Therefore revised time-related cost following a 6-week reduction in the contract period =

$$\frac{£1{,}868.00 \times 46 \text{ weeks}}{52 \text{ weeks}} = £1{,}652.46$$

Step 3. Add back unchanged fixed cost element:

= £1,652.46 + £306.00
= £1,958.46

Step 4. Add back relevant percentage for overheads and profit:

= £1,958.46 × 1.15
= £2,252.23

(b) Temporary telephone

Step 1. Deduct for overheads and profit, say 15 per cent.

$$\text{Therefore net cost} = \frac{£2{,}000.00}{1.15} = £1{,}739.13$$

Step 2. Isolate the fixed costs from the time-related costs:

Fixed costs: installation of system.
Time-related costs: quarterly rental charge, cost of calls (not strictly time-related).

Assume costs as follows:

Installation cost	£75.00 per week
Quarterly rental	£52.00
Cost of calls, say	£28.00 per week

Total time-related costs = £1,739.13 − £75.00
= £1,664.13

Therefore revised time-related cost following a 6-week reduction in the contract period =

$$\frac{£1{,}664.13 \times 46 \text{ weeks}}{52 \text{ weeks}} = £1{,}472.12$$

Step 3. Add back unchanged fixed-cost element:

= £1,472.12 + £75.00
= £1,547.12

Step 4. Add back relevant percentage for overheads and profit:

= £1,547.12 × 1.15
= £1,779.19

(c) Water for the works

Step 1. Deduct for overheads and profit, say, 15 per cent:

$$= \frac{£1{,}500.00}{1.15} = £1{,}304.35$$

In this case, there is no time-related element, but nevertheless an adjustment may still be applicable since part of the cost of this item is directly related to the contract value. (Most water authorities charge for building water in this way irrespective of the actual amount used.) Thus if the contract sum alters, as it will do following the adjustment to the Preliminaries, there should, in theory at least, be a small revision to make here.

Step 2. Isolate the fixed costs from the value-related costs:

Fixed costs: time and materials in installing temporary supply to standpipes, washbasins, etc. and dismantling on completion.

Assume:

1 plumber for 3 days at £5.00 per hour	=	£120.00
1 labourer for 2 days assisting at £4.00 per hour	=	64.00
Hire of excavator and driver for digging and backfilling trenches at £12.00 per hour for ½ day	=	48.00
Polythene tubing and fittings, say, £150.00		150.00
	=	£382.00

Therefore value-related cost = £1,304.35 − £382.00
= £922.35

Assume contract value is reduced by, say, 10 per cent:

Therefore value-related cost = £922.35 – 10%
= £830.11

Step 3. Add back fixed-cost element:

= £830.11 + £372.00
= £1,202.11

Step 4. Add back relevant percentage for overheads and profit:

= £1,202.11 × 1.15
= £1,382.43

(d) Non-productive overtime

Whereas the above three items have been reduced in value following the 6-week reduction in the contract period, it is likely that the cost of this item will **increase** since the same amount of work now has to be done in a shorter time, probably necessitating the use of more overtime working.

Step 1. *Assume*: Number of overtime hours needed.
Assume: An average of 10 men (7 skilled, 3 unskilled) would have been needed for the 6 weeks. These 60 man weeks now have to be worked as overtime during the reduced contract period.

42 skilled man weeks = 42 × 39 skilled hours = 1,638 hrs

18 unskilled man weeks = 18 × 39 unskilled hours = 702 hrs

Normal skilled rate = $\frac{£102.37\frac{1}{2}}{39}$ = £2.63 per hour

Overtime rate
= $\frac{£88.33\frac{1}{2}\text{ (basic)}}{39}$ × $1\frac{1}{2}$ (at time-and-a-half)
= £3.40 per hour

Therefore **extra over** cost per skilled overtime hour
= £3.40 – £2.63 = 0.77
Add approx. 50% for on costs 0.39
£1.16

Therefore extra cost of skilled overtime working
= 1,638 hrs × £1.16
= £1900.08

Normal unskilled rate = $\frac{£87.16\frac{1}{2}}{39}$ = £2.24 per hour

Overtime rate
= $\frac{£75.27\text{ (basic)}}{39}$ × $1\frac{1}{2}$ (at time-and-a-half)
= £2.90 per hour

Therefore extra over cost per unskilled overtime hour
= £2.90 – £2.24 = 0.66
Add approx. 50% for on costs = 0.33
£0.99

Therefore extra cost of unskilled overtime working
= 702 hrs × 0.99
= £694.98

Total net extra cost of overtime = £1900.08 + 694.98
= £2,595.06

Step 2. Add overheads and profit at, say, 15 per cent:
= £2,595.06 × 1.15
= £2,984.32

Step 3. **Add** to original bill rate:
= £3,000.00 + £2,984.32
= £5,984.32

Index